The Best Advanced

Paper Airplane COMPENDIUM

Carmel D. Morris

First published by Perigee, an imprint of Putnam New York, as *The Best Paper Aircraft*.

This edition published by Flight Time, an imprint of Gryphon Chess, 2015.

ISBN: 9780994315328
Cover design: Cloud Bunny
Line drawings: Cloud Bunny
Typeset in The Mix 5-Regular and Agency FB
Copyright: Carmel D. Morris
NLA CiP entry – Morris, Carmel, date-
This edition is printed in the United States of America.
10 9 8 7 6 5 4 3 2 1

Preface

Arrow Glider

Welcome to this best-of-the-best edition. This book is released to complement the new Apple® and Android™ app, *Paper Aircraft Advanced* which is available for free download.

Although a hard choice by our editors, this book contains 27 models, the best chosen from the three best-selling *Best Advanced Paper Aircraft* books, along with tips and tricks to help you master folding techniques and improve flying skills.

This book is your single source from which to fold what is considered by many experts to be the best paper airplane designs in the world. These original designs have been used extensively in advertising, by the BBC, for inflight magazines, for 3D-printed hard-form modelling, paper airplane competitions, by other authors inspired by this series, and much more.

Folding diagrams are easy-to-follow and basic folds are repeated so you do not need to page back anywhere. All you need is a sheet of A4 or Letter paper to get started.

Some models are more difficult than others, hence the 'advanced' aspect of this book; but if you like a challenge, this book is ideal – and much better training for young flight engineers than those pretty tear-out books that don't last. Some of the more complicated models come with photos that show more detail, so don't worry if you get stuck; you won't have to call your Mom or Dad, unless they want to join in. Sure my grand folks love paper airplanes!

When it comes to throwing and flight theory, information is provided at the end of this book to help you throw your masterpieces out of the park and beyond.

Happy folding and flying!

Col. Dwight Edwards (Retired)

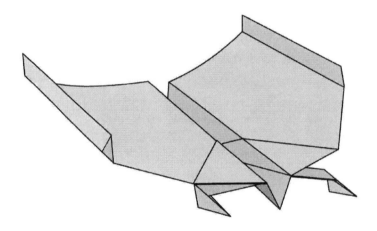

Dive Bomber MK1

Contents

Introduction

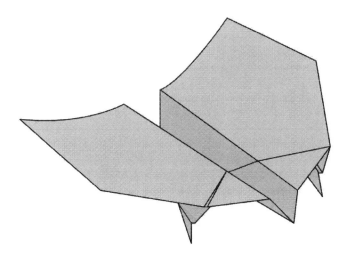

Glider with landing gear

Welcome to the fun world of paper airplanes. Not just your average dart, this book will show you how to fold and fly some amazing models. Most of the models in this book are folded; you do not have to cut anywhere but for rear tail lift in some models, shaping a wing section in the WWI Fokker and cutting a notch for the Elasto shooter.

Paper Sizes

All models in this book are made from standard US Letter which also works well for A4, despite a slightly different aspect ratio. Photocopier paper is ideal.

Unlike some other books, this book is not dependent on a limited number of page tear-outs; in fact we encourage you to use ordinary paper, work with these models, and be inspired to design your own.

Folding

In this new edition, base folds are repeated for each model that uses them. This should provide for near-seamless reading as you do not need to return to an earlier section.

Photos are also provided for models that have difficult folds.

Always remember safety; do not throw your darts in the direction of people. Pointed objects could hit an eye.

Symbols and Folding Techniques

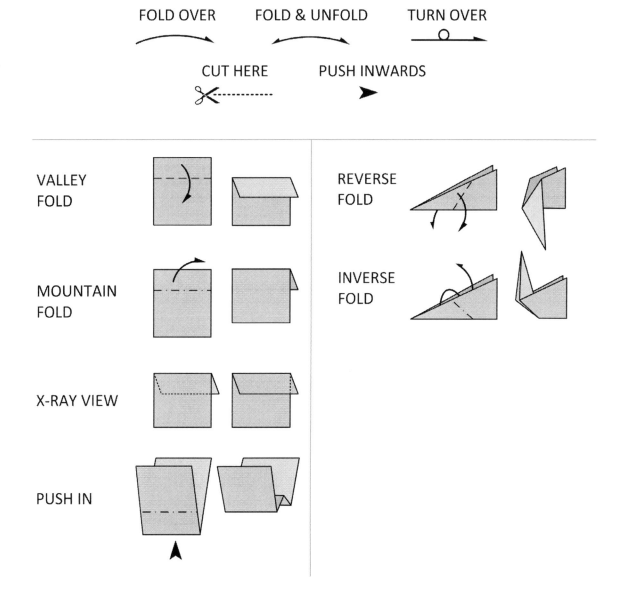

FOLD OVER   FOLD & UNFOLD   TURN OVER

CUT HERE   PUSH INWARDS

VALLEY FOLD

MOUNTAIN FOLD

X-RAY VIEW

PUSH IN

REVERSE FOLD

INVERSE FOLD

Rabbit Ears

Rabbit ears are common origami techniques that are also used in some paper airplane designs.

Make a diagonal crease-fold, and then two intersecting crease-folds. Bring in the sides and pinch together to form a point, and then flatten the point.

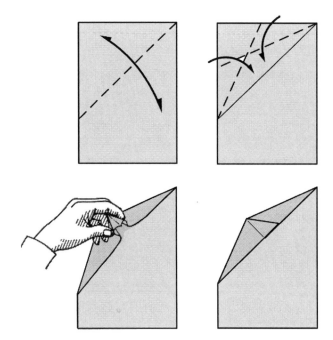

Elasto Kinetic Jet

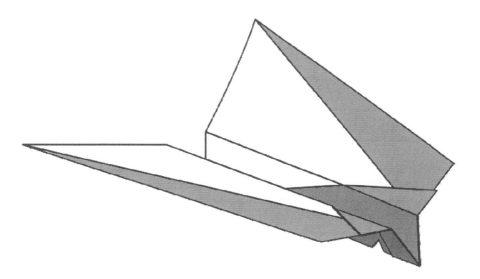

While your buddies are haphazardly throwing their paper darts around the schoolroom, beat them to the top of the class with this rubber-band-powered, target-acquiring jet missile.

1 Elasto Kinetic Jet

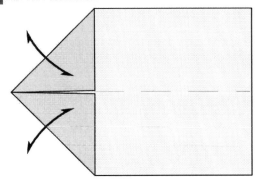

Start with paper crease-folded lengthwise in half and the left corners folded in as for a regular dart. Now unfold those corners.

2 Elasto Kinetic Jet

Fold corners to meet the creases made in Step 1.

3 Elasto Kinetic Jet

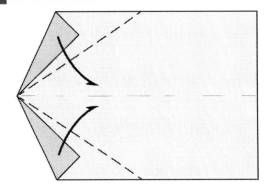

Fold the leading edges to meet the center crease.

4 Elasto Kinetic Jet

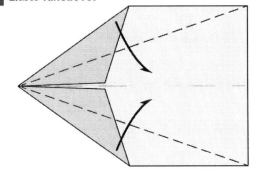

Fold again to the center crease.

5 **Elasto Kinetic Jet**

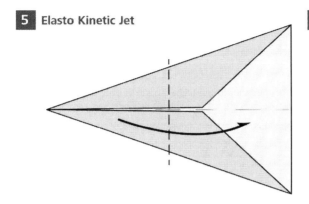

Fold left point to the right and make sure it goes past the right edge by a quarter inch or so.

6 **Elasto Kinetic Jet**

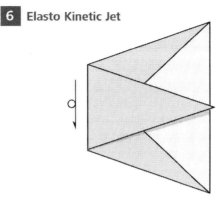

You can see the rigid fuselage we are making. This will take the strain when launching. Turn the model over.

7 **Elasto Kinetic Jet**

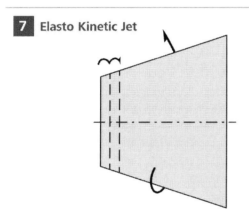

Fold left edge to the right two times at roughly half an inch each fold, and then fold in half.

8 **Elasto Kinetic Jet**

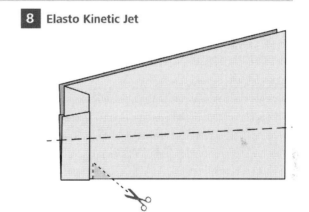

Using box cutter or scissors, cut a notch behind the folded leading edge, and then fold down the wings.

9 Elasto Kinetic Jet

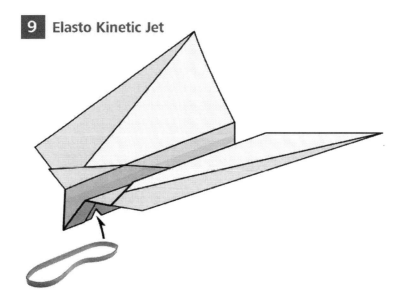

Your completed jet is ready to launch. Hook an elastic/rubber band into the cut-out notch and loop the other end around your thumb. With your other hand hold the center of the fuselage, aim high, stretch back and release. It flies darn fast!

Smooth Flyer

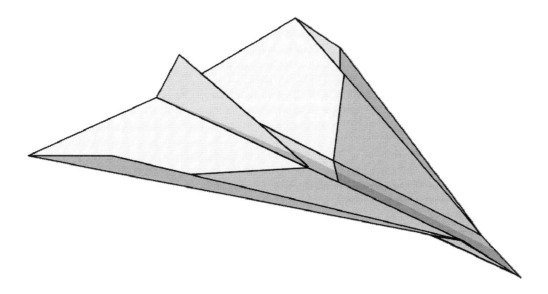

This craft offers stability and simplicity with little drag; even the teach will be impressed when you fling this across the schoolroom.

1 Smooth Flyer

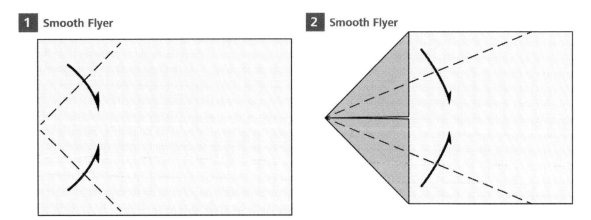

Fold corners as for a regular dart.

2 Smooth Flyer

Fold corners again.

3 Smooth Flyer

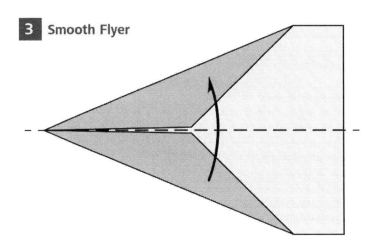

Fold in half.

4 Smooth Flyer

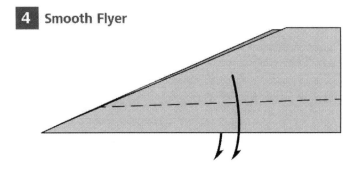

Fold wings down at a shallow angle.

5 Smooth Flyer

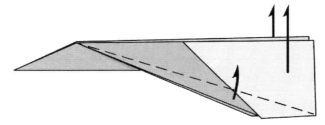

Angle side fins from nose to tail and then swing the wings up.

6 Smooth Flyer

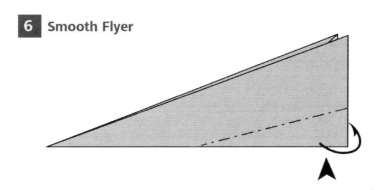

Crease well and create a tail by pushing in the bottom
of the fuselage where it meets the wing crease.

7 **Smooth Flyer**

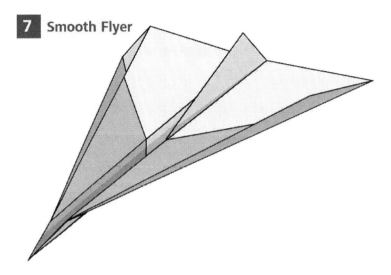

And that's it; a very easy aircraft to build that looks cool too. Hold at a level angle and throw with slight force. If she strays, curl the tail section slightly.

Arrow Dart

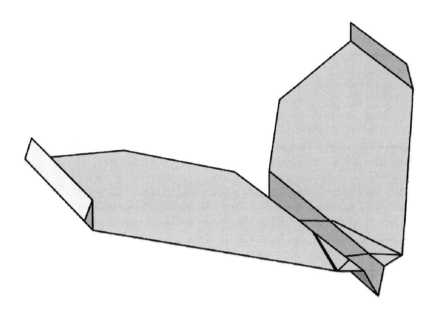

On casual inspection, you could be fooled into thinking this craft has a cut-out wing shape but actually this is folded using one piece of paper with no cuts. This craft, when folded to as exacting symmetry as possible, will glide with a shallow descent; ideal for long distances, and often carried away on a breeze.

If Steps 2 – 4 seem difficult, don't worry; there are some photo tips at the end of the instructions for this craft.

1 Arrow Dart

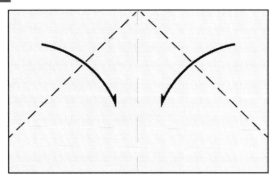

With the paper facing you in landscape orientation, fold in the corners.

2 Arrow Dart

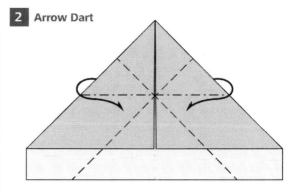

First crease-fold these sections and then bring the sides inwards, bringing the top point downwards.

3 Arrow Dart

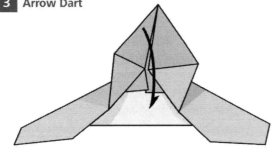

Almost there... This is similar behavior to making a base fold as for Super Looper.

4 Arrow Dart

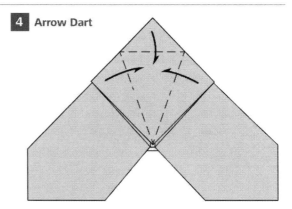

Fold the sides of the upper flaps to the center; and lock their corners into the top section as you fold that down.

5 Arrow Dart

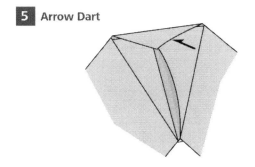

6 Arrow Dart

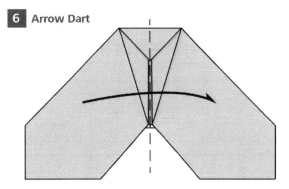

Step 4 almost done. Lock the side corners into the top section.

Fold the sides of the upper flaps to the center; and lock their corners into the top section as you fold that down.

7 Arrow Dart

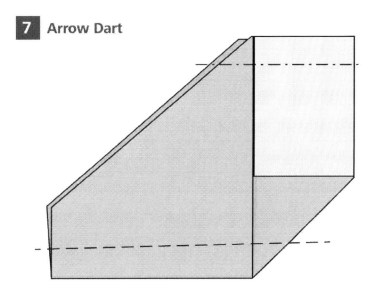

Fold down the wings and crease well; and then fold the side fins.

8 Arrow Dart

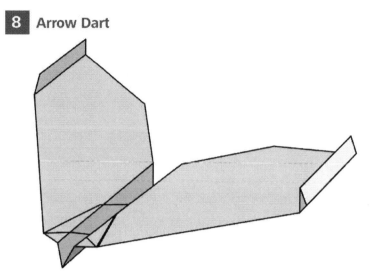

Your finished Arrow Dart. Throw high and against the wind.

If your craft stalls or suddenly dives, curl up the tail section slightly. It's important to make an identical curl on each side; otherwise the craft could spin and crash.

You will notice if you gently throw indoors (no wind), the glide descent is shallow. When folded noting strict symmetry, this craft will glide well and far outside, and stay in the air for longer periods. Experiment with different paper sizes and see how differently it can perform.

Difficult Folds in Detail

If you have trouble folding this craft at Steps 2 – 4, the following photos will help.

Crease-fold first diagonal.

Make sure the fold is well-creased. Unfold and fold opposite diagonal.

The opposite diagonal intersects the center vertical crease and other diagonal crease you just made.

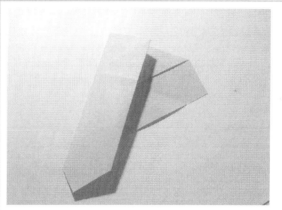

Make sure the fold is well-creased and then unfold.

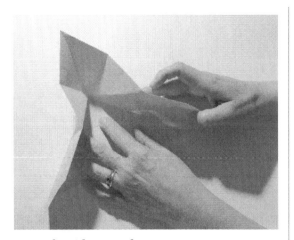

Bring the sides together.

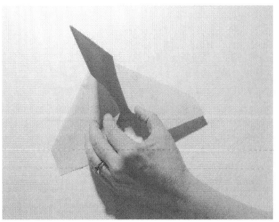

Hold bottom of new flaps together, open out in the middle and flatten down.

Basic fold complete.

Long Distance Glider MKI

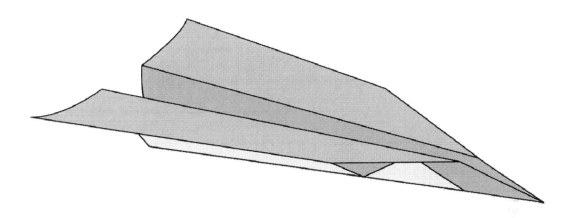

If you are sitting at the back of a lecture theater or cinema and do not find it entertaining, this dart will get your message across!

1 Long Distance MK1

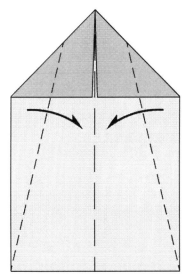

Begin with a sheet of paper crease-folded in half and the top corners folded in. Fold the side corners in to meet the center crease.

2 Long Distance MK1

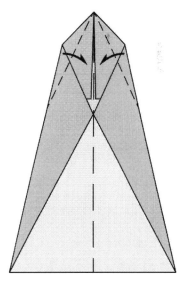

Fold the corners in to meet the center crease.

3 Long Distance MK1

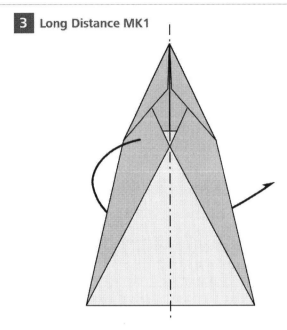

Fold behind lengthwise in half.

4 Long Distance MK1

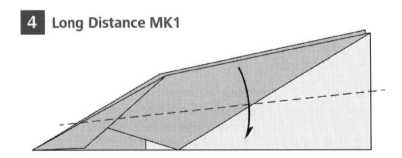

Fold the wings down and then level them at a right angle to the fuselage.

5 Long Distance MK1

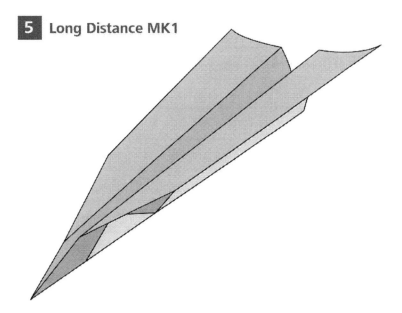

Your dart is now complete. Curl the tail section slightly upwards for lift if you need it. If the model stalls, fold the nose by an inch or so back into the fuselage.

Throw with gentle force at approximately 30 to 40 degrees in an upward direction. Because of its length, this dart will prove accurate in meeting its target.

Super Wing

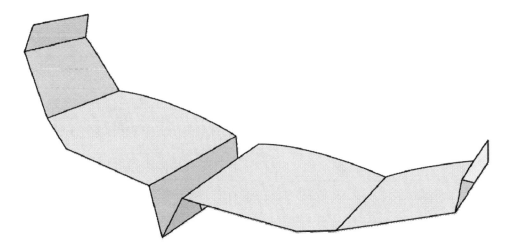

This incredible paper wing is surprisingly stable as a glider and can perform some aerial manoeuvres too. If you tape the fuselage together you can throw it harder and watch it catch some breezes.

We are currently building a quad-copter paper plane launcher to launch gliders like this that can hover nicely on upper breezes. One of the things you can do is design your own remote powered tilt platform to mount on a standard quad-copter, just using a lunchbox lid, for example.

On ours, using Arduino, we raise the quad-copter as high as possible and then remotely set a stepper motor to tilt a platform at one end, thus creating a ramp. The paper plane sitting on it slides off and is instantly launched at a height.

The design possibilities are endless —you could drop water bombs instead, if you want!

1 Super Wing

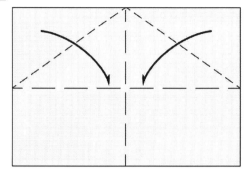

Begin with a sheet of paper facing you horizontally and crease-folded in half in both directions along the dashes. Fold the top corners in.

2 Super Wing

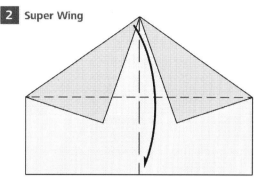

Fold the corners in to meet the center crease.

3 Super Wing

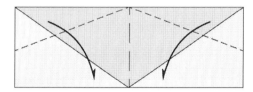

Fold corners to meet bottom edge.

4 Super Wing

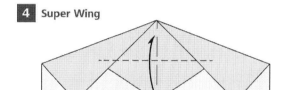

Fold lower point up to nose, tucking under the upper flaps.

5 Super Wing

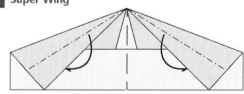

Fold back the flap edges along the dots and dashes (mountain fold) to strengthen the leading edge.

6 Super Wing

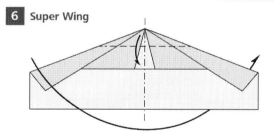

Fold nose down, and then fold behind in half. Rotate model to face you as seen in Step 7.

7 Super Wing

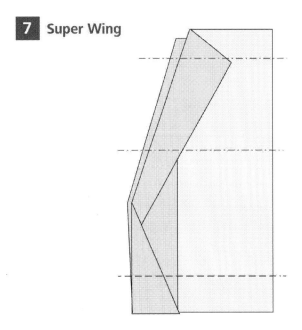

Where the valley fold is indicated, fold the wings down; where the mountain folds are indicated, fold wing fins back.

8 Super Wing

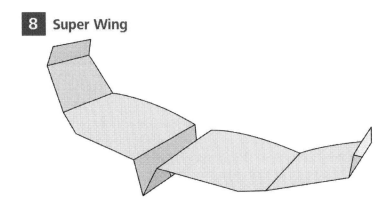

The completed Super Wing. Throw horizontally with slight force. If your model is perfectly symmetrical, it should be able to perform loops when thrown with greater force in an upward 45 degree angle.

To make the Super Wing return to you in a relative horizontal circle, throw it upwards and away from you at approximately 30 degrees with its underside facing you.

Hang Glider

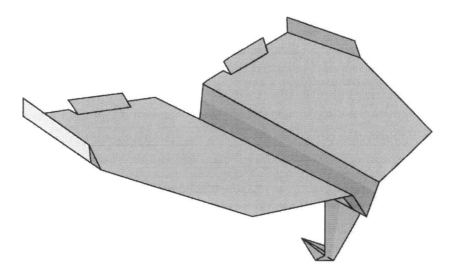

An unusual glider that has a 'winged keel', it is surprisingly stable and can catch those sea breezes to carry it to who knows where? The strong forward section means you can throw it hard and high and the keel section will make the glider level out every time.

1 Hang Glider

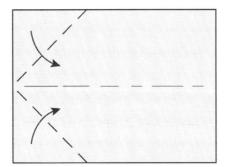

Crease-fold in half, and then fold the corners in.

2 Hang Glider

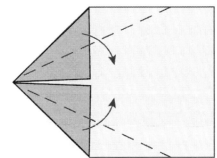

Fold the sides in as for a regular glider.

3 Hang Glider

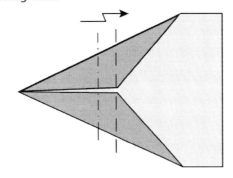

Make a 'stair-step': fold left point across to the right and then back to the left by an inch or so.

4 Hang Glider

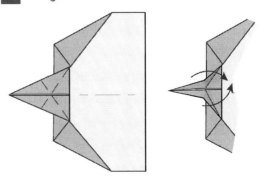

Make a rabbit ear fold; crease well and pinch the sides together.

5 Hang Glider

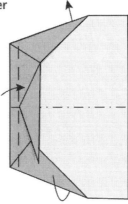

Fold over the leading edge and then fold in half; allow the 'keel' to swing out (do not fold it).

6 Hang Glider

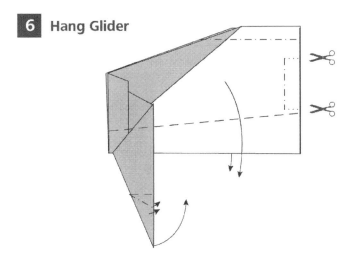

Fold wings, side fins, and snip trailing edges for lift. On the keel section, crease horizontal and angled folds and tuck your finger inside, and then lift to make a small horizontal 'wing' section that is flattened and parallel to the fuselage.

7 Hang Glider

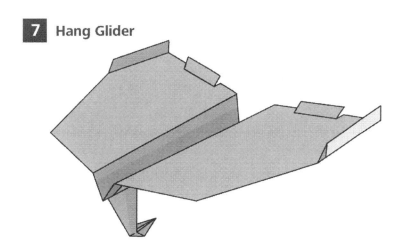

Throw at any angle and the keel will help to right the craft, acting as ballast. Experiment with the keel section by making it larger to see how the flight performance changes.

If you do not cut the rear flaps, the craft will swing from side to side as it glides, which makes for interesting dive-bombing behavior amidst throwing crossfire from others; such as paper darts, water bombs —or key lime pies!

Difficult Folds in Detail

If you have trouble folding this craft at Steps 4 – 5, the following photos will help.

This is the stair-step fold we made in Step 3.

Fold a diagonal to the right, and then unfold.

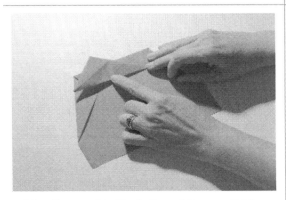

Fold a diagonal to the left, and then unfold

Pinch the sides together to form a point.

Almost there...

Flatten the point to one side.

Long Distance Glider MKII

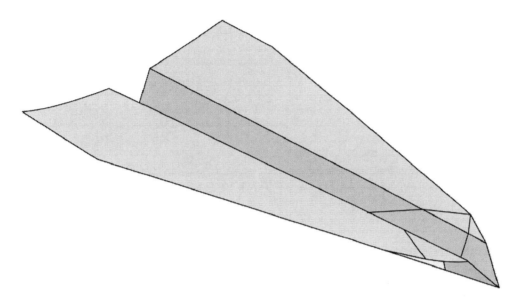

Here is a dart that can be thrown with greater force due to its solidly-constructed split-nose section.

While staying at the Hotel del Coronado in San Diego, we snuck into the red-roofed turret to have a look. On the lookout above we were able to throw these and they stayed aloft, sailing across the ocean air, coming to land (we presumed) at the naval base next door, as it had flown off in that direction. We hope they made a good touchdown on a carrier!

1

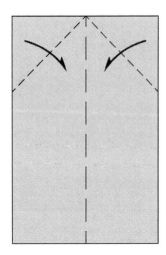

Crease-fold in half, and then fold the corners in.

2 Long Distance MK2

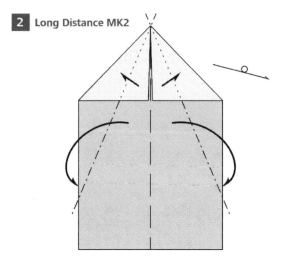

Fold the sides behind, flipping out the top corners. Next, turn the paper over.

3 Long Distance MK2

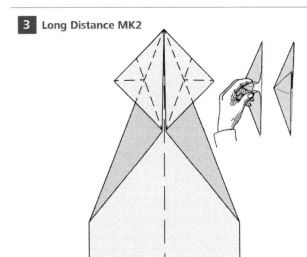

Make 'rabbit ears' on the top sections by crease-folding and then pinching the edges together to make an 'ear' section on each side (this is shown in more detail at the end of the instructions for this model).

4 Long Distance MK2

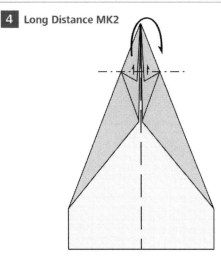

Fold the nose behind and flip forward the two bunny ears.

5 Long Distance MK2

6 Long Distance MK2

7 Long Distance MK2

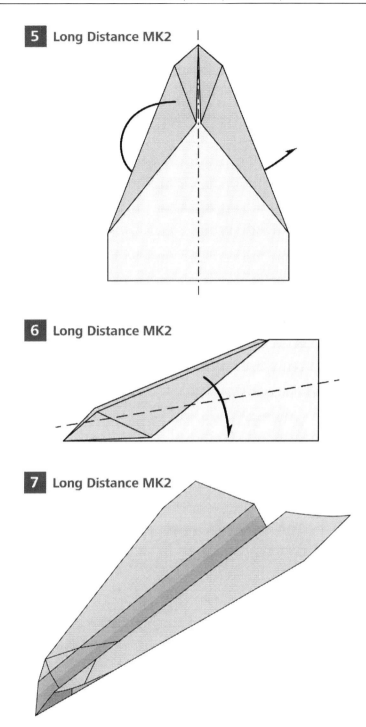

The completed glider; curl up the tail section slightly if you need extra lift. Throw upwards with moderate force, preferably against or slightly side-on to the wind.

This model is ideal if you are making 3D-printed launcher guns or launch ramps from quad-copters, etc.

Difficult Folds in Detail

If you have trouble folding the rabbit ears for this craft at Steps 2 – 4, the following photos will help.

At Step 2, with the model upside down, fold the sides in and flip out the flaps underneath (do not fold them).

You can see the small triangle section flipped about.

Make sure the underlying fold is well-creased.

Make a diagonal fold on the small triangle section, and then unfold.

Make another diagonal fold in the opposite direction and unfold.

Pinch the sides together to make the rabbit ear.

Flatten and have the point folded forward.

Repeat for the left side.

Rabbit ears complete.

Super Looper

Many authors have tried to copy this model and have not succeeded! This is the original and best design for outstanding looping performance.

This craft begins with a standard basic fold. In step 1 you will see diagonals (valley folds). The horizontal dot-dash (mountain fold) line means you crease behind. This allows you to bring the sides together and collapse the paper to make the basic fold.

1 Super Looper

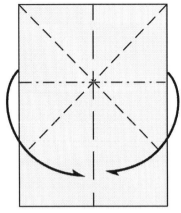

Make all creases (the horizontal is folded behind) and then unfold. Now bring in the sides and collapse down the top as shown in Steps 2 and 3.

2 Super Looper

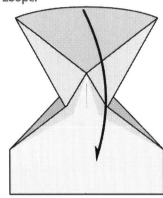

Almost there...

3 Super Looper

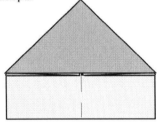

The completed basic fold. If you are struggling with this fold, photos are provided after instructions for this model, to describe the base fold procedure in more detail.

4 Super Looper

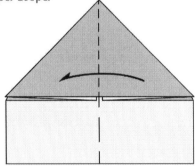

Larger view; swing the upper right flap to the left.

5 Super Looper

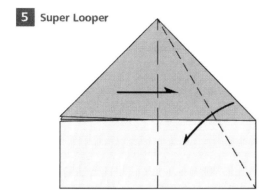

Fold the side in from nose to bottom right corner. Next, flip back to the right the upper left flap. Repeat Steps 4 and 5 for the other side.

6 Super Looper

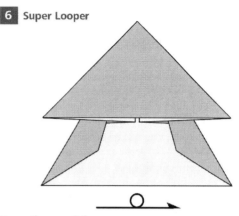

Turn the model over.

7 Super Looper

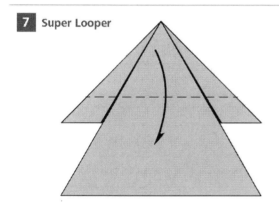

Fold nose down.

8 Super Looper

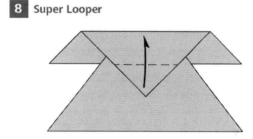

Fold point back up; this will create solid fuselage for you to grip for throwing.

9 Super Looper

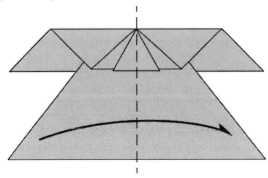

Fold in half and swing the model about to face you for Step 10.

10 Super Looper

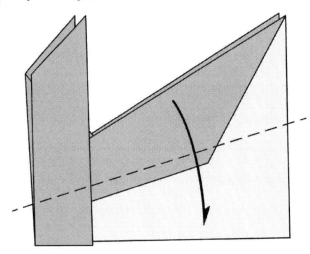

Fold down wings, and keep them relatively level.

11 Super Looper

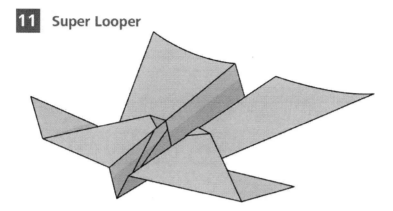

The completed Super Looper; throw upwards with force and it will do a massive loop. If thrown at the oncoming wind, it will be picked up and carried away. Adjust the amount of tail curl (more trailing edge upward curl for indoors, meaning a smaller diameter loop).

Difficult Folds in Detail

If you had trouble with the base fold for this craft at Steps 1 – 3, the following photos will help.

Having made two diagonal valley creases, make the horizontal mountain crease; push your finger in the middle and let the sides 'pop up'.

Paper has 'popped up'. Now bring the sides together.

Almost there...

With the sides brought together, swing the curved section down and flatten it.

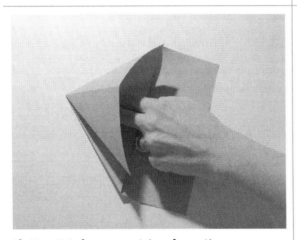

Flatten it to become a triangle section.

Completed base fold.

WWI Fokker Eindecker

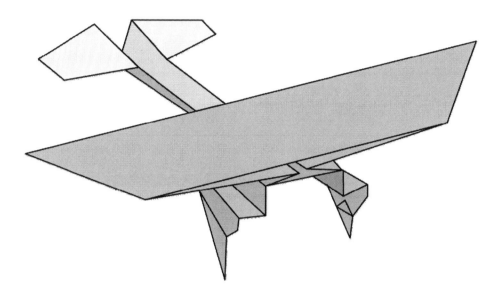

This early monoplane (1915) is not only a novelty, but flyable! If the lesson is Aviation History, show your appreciation by landing one of these on the teacher's desk.

It is recommended to snip an inch off the side of the paper before starting, so that the sheet has a longer aspect ratio. While the model flies well using A4 or Letter paper, a longer fuselage will provide better stability.

1 WWI Fokker Eindecker

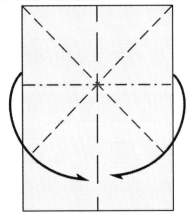

In Step 1, 2 and 3, crease and make a basic fold, same as for the Super Looper.

2 WWI Fokker Eindecker

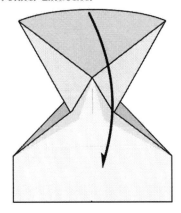

Almost there...

3 WWI Fokker Eindecker

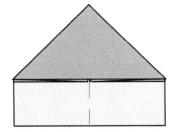

The completed basic fold.

4 WWI Fokker Eindecker

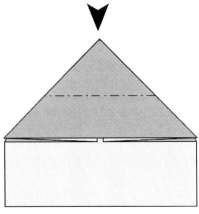

Crease well on all sides and push the point inwards. This means opening out the fold, pushing in the center, and then collapsing it all again.

5 WWI Fokker Eindecker

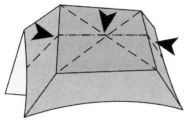

Almost there... Flatten the fold again by pushing against the creases in the left and right sides. It should collapse nicely.

6 WWI Fokker Eindecker

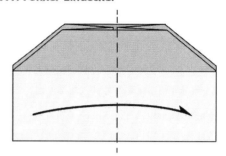

Fold in half the entire model.

7 WWI Fokker Eindecker

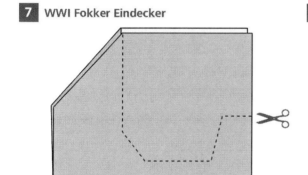

This is one of the few models in this book that requires cutting – but it's worth it! Cut this area off.

8 WWI Fokker Eindecker

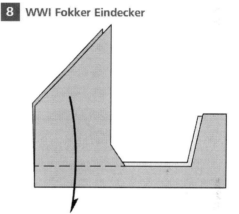

Fold one wing side down. In the following steps we will make the undercarriage.

9 WWI Fokker Eindecker

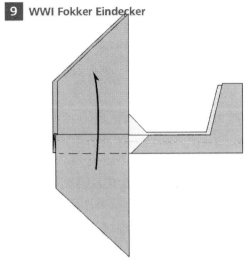

Fold back up along bottom edge of fuselage.

10 WWI Fokker Eindecker

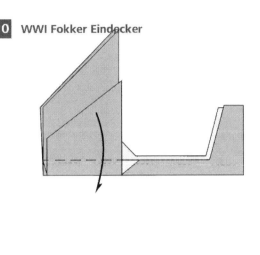

Fold down again.

11 WWI Fokker Eindecker

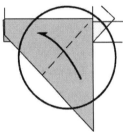

Steps 11 – 14 show a cutaway section detailing the landing gear assembly for one side. When done, repeat for the other side.

Fold flap diagonally up to the left.

12 WWI Fokker Eindecker

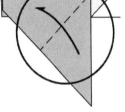

Fold edge down.

13 WWI Fokker Eindecker

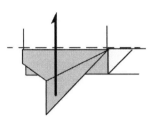

Swing this section up.

14 WWI Fokker Eindecker

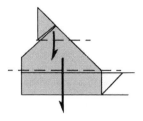

Fold upper triangle to lock the leg fold and then swing the landing section back down.

15 WWI Fokker Eindecker

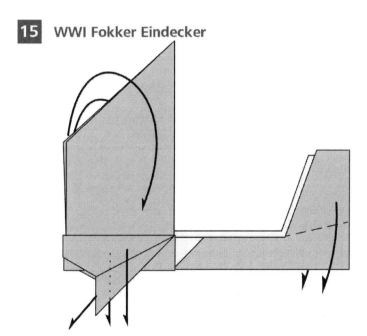

Bring the wings down, grasping each end at the tips and pulling gently to straighten the wingspan. Lower the undercarriage to a position which supports the plane. Fold tail wings down and you are now ready to fly.

16 **WWI Fokker Eindecker**

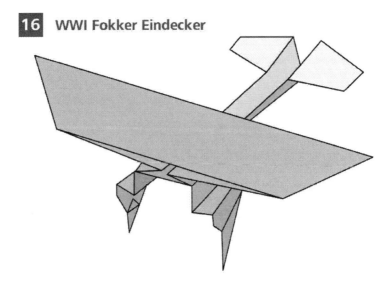

This finished WWI craft is a gentler flier than the original (which turned the British aircraft into 'Fokker fodder' during World War 1).

Place your index finger inside the fuselage at the tail section while grasping the underside of the fuselage with thumb and forefinger. With a slight downward movement, gently let go of your craft. It will glide to a soft landing.

Glider with Landing Gear

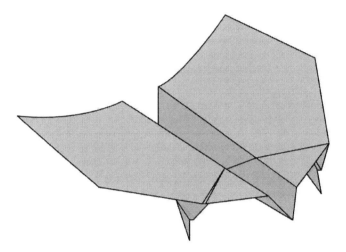

Looks good when it makes a landing on a smooth surface such as a desk or polished floor, and the undercarriage also provides stability during flight. The undercarriage is more refined than the WWI aircraft and can be used in that plane also.

The craft uses the same basic fold as the Super Looper, and the base fold steps are repeated for your convenience.

1 Glider with Landing Gear

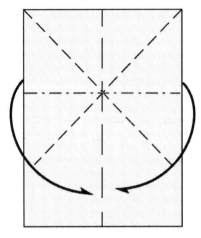

In Step 1, 2 and 3, crease and make a basic fold, same as for the Super Looper.

2 Glider with Landing Gear

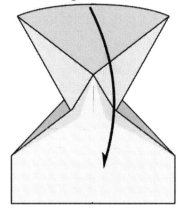

Almost there...

3 Glider with Landing Gear

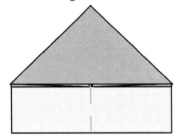

The finished basic fold.

4 Glider with Landing Gear

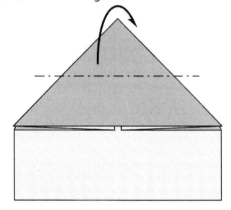

Larger view; fold nose section behind.

5 | Glider with Landing Gear

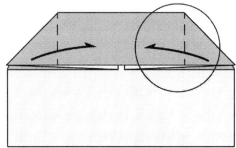

In Steps 5 – 12 we will make the undercarriage. First,
fold upper flap corners in towards center.

6 | Glider with Landing Gear

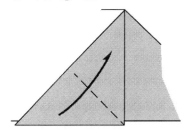

This is a close-up view of the undercarriage on
the right-hand side. Fold up to the right.

7 | Glider with Landing Gear

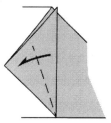

Fold back to the left.

8 | Glider with Landing Gear

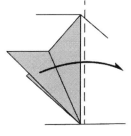

Swing this section to the right.

9 | Glider with Landing Gear

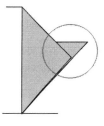

Let's take a closer look at the small triangle
section...

10 Glider with Landing Gear

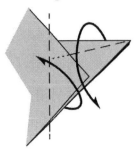

Step 10 shows a close-up cutaway of the landing feet detail. As you fold the triangle section to the left, fold the upper horizontal edge down.

11 Glider with Landing Gear

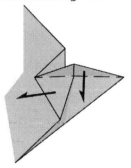

Almost done; as you fold to the left, flatten to upper section down.

12 Glider with Landing Gear

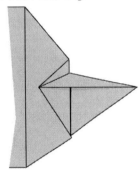

Done! Repeat for the other 'wheel'. If your glider tips over on landing, round-out the 'wheel' by folding the right point to the left by a quarter inch or so.

13 Glider with Landing Gear

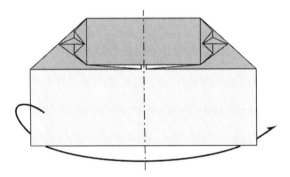

Almost done; as you fold to the left, flatten to upper section down.

14 Glider with Landing Gear

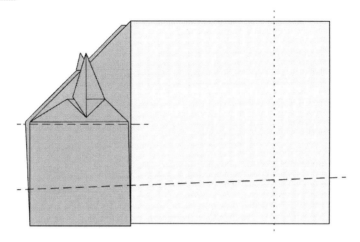

In Step 14, fold wings and undercarriage down. If your glider stalls, cut the trailing edge back around half an inch.

15 Glider with Landing Gear

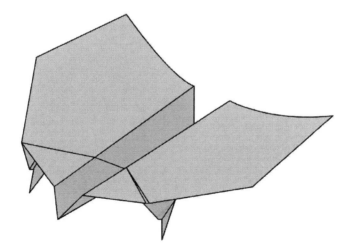

Ready to fly! Throw horizontally away from you towards a smooth surface and watch it land gracefully.

Difficult Folds in Detail

If you have trouble folding the undercarriage at Steps 10 – 12, the following photos will help.

The small triangle section needs to be folded to the left, and when you do so, the upper edge behind it is folded down to lock the undercarriage in place.

Fold carefully to the left.

This is the paper rotated with the front of the leg facing us.

Close-up; as you fold across, the bent edge (center of image) is folded flat.

Almost there...you can see the reinforcing edge taking shape. Flatten this edge and then swing the paper about.

You can see how the leading edge of the undercarriage is folded back to lock the landing gear in place; this offers reinforcement for landing.

Landing leg complete. Repeat for the other side.

Landing Tips

This craft is designed for landing on smooth surfaces such as a polished floor, or a desktop. If the craft tips over on landing, fold the leg points back by a quarter inch or so, or snip off.

You could also modify the landing legs by turning them into 'guns' or 'skis' by reverse-folding the legs forwards.

Dive Bomber MKI

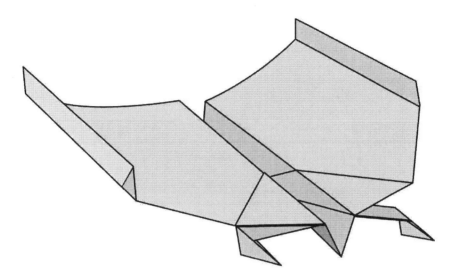

This craft makes for a neat dive bomber when thrown hard and downwards towards a target.

Also known as a sea plane (though there is a better one in the *Paper Boats!* book), this craft can actually float somewhat in water if the undercarriage is bowed out and the craft is folded using plastic sheeting/wrapping paper.

This model uses the same basic fold as the Super Looper, and base fold steps are repeated here for your convenience.

1 Dive Bomber 1

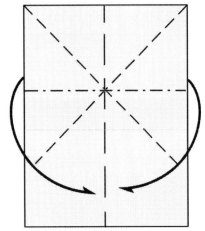

Crease well the diagonal valley folds and the horizontal mountain fold. Bring the sides together, and top down.

2 Dive Bomber 1

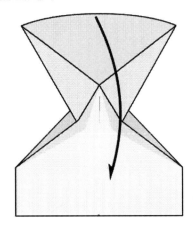

Almost there...

3 Dive Bomber 1

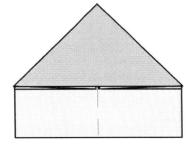

Basic fold is complete.

4 Dive Bomber 1

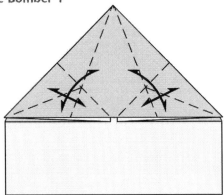

Larger view; make all the creases and pinch the sides together to make two big bunny ears, by following Steps 5 – 7.

5 Dive Bomber 1

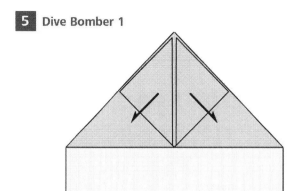

Fold and unfold...

6 Dive Bomber 1

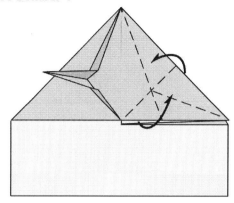

Left side almost complete; flatten and repeat for right-hand side.

7 Dive Bomber 1

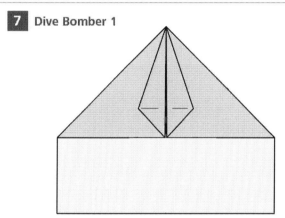

How it should look.

8 Dive Bomber 1

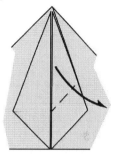

Make a diagonal fold on the right-hand point.

9 Dive Bomber 1

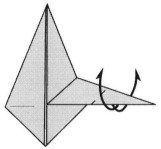

Make a diagonal crease, put your finger inside the flap and open it inside-out).

10 Dive Bomber 1

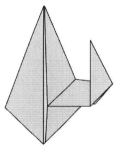

How it should look; repeat for the left side.

11 Dive Bomber 1

12 Dive Bomber 1

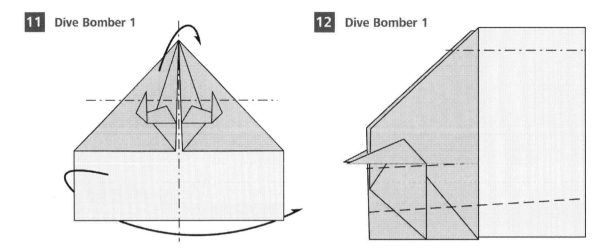

Fold nose section behind and then fold in half behind.

Fold down the 'guns' with the wings; otherwise fold them downwards and open out the legs to make skis if you want to make a 'water landing' craft.

13 Dive Bomber 1

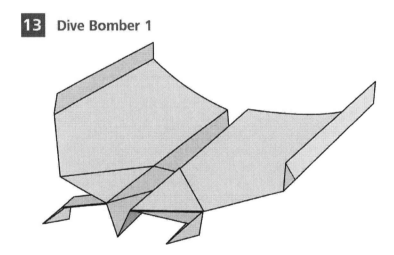

The completed diver bomber/sea plane; if the craft stalls, trim off the trailing edge wing section by around half an inch.

Runway Skimmer

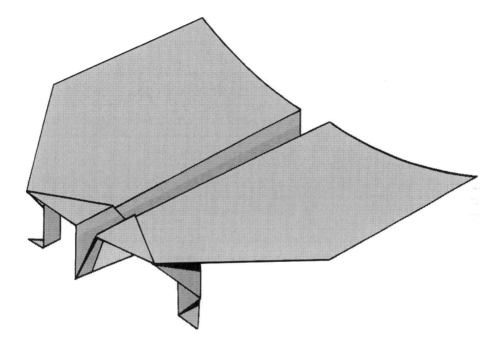

This is a simple craft with flat landing feet for easy skimming across a desk or other smooth surface. As it lands, it skims and keeps flying, and is ideal for landing competitions.

1 Runway Skimmer

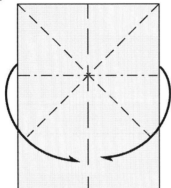

Make the intersecting diagonal creases (shown in the following three steps), plus a horizontal one that also intersects the diagonals. You will then be bringing the sides together to form the same basic fold as for the Super Looper.

2 Runway Skimmer

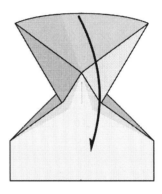

Almost there... Bring the top edge down and then flatten the fold.

3 Runway Skimmer

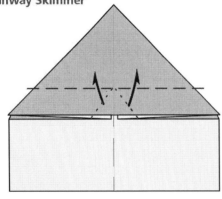

Larger view. Crease along the dashes, place fingers under upper flap of upper triangle section and lift up the edge.

4 Runway Skimmer

Almost there... This image shows a cutaway of the center section. Flatten, ready to make landing legs.

5 Runway Skimmer

6 Runway Skimmer

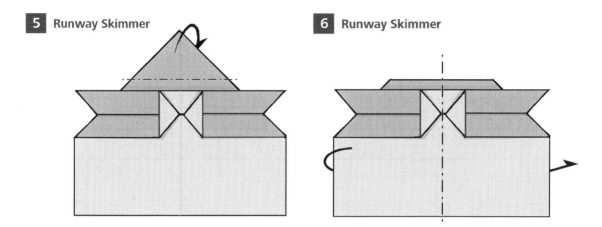

Fold the nose back.

Fold in half behind and swing the model about as seen in Step 7.

7 Runway Skimmer

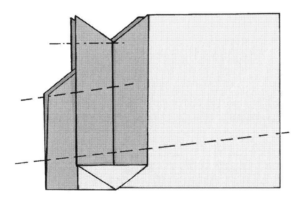

Fold down wings and legs parallel to each other, and fold the feet in the opposite direction.

8 Runway Skimmer

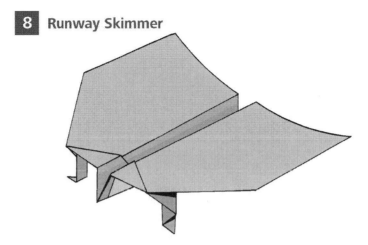

You're ready to fly; throw with a shallow descent towards a smooth surface such as a polished floor, desktop, kitchen bench top etc., the spring action from the undercarriage will lift the craft enough to skim along the surface.

This craft has also been used for the Aircraft Carrier Game; using the origami aircraft carrier in the book, *Paper Boats!*, kids and adults make a big aircraft carrier for the swimming pool and try to land their paper airplanes on it.

Aircraft Carrier from Paper Boats!

Dive Bomber MKII

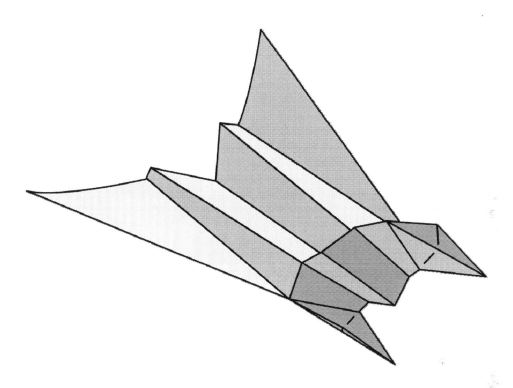

This airplane makes huge swooping dives, picking up momentum while gathering lift. The fuselage can also be sealed for harder throws. This is a great airplane to experiment with.

This model uses a different base fold from that described in the Super Looper. To get to the image shown in Step 1, have a piece of paper with a center lengthwise crease and the upper corners folded in as for a regular paper dart, and then those corners unfolded and re-folded to meet the new crease. This is the same as Steps 1 and 2 of the Elasto Kinetic Jet.

If you have trouble making the basic fold, there is a set of photos following the instructions for this model that provides more detail.

1 Dive Bomber 2

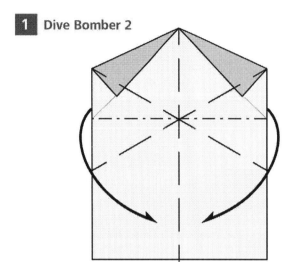

Make a horizontal mountain fold and two intersecting diagonal creases to create an upper section similar to that shown in the Runway Skimmer. This procedure is shown in the following steps.

2 Dive Bomber 2

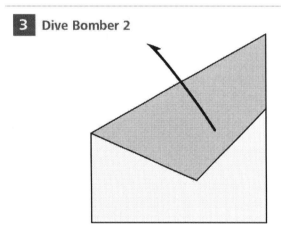

Upper right corner folded to the left, then unfold.

3 Dive Bomber 2

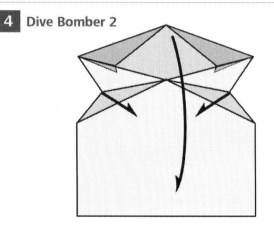

Upper left corner folded to the right, then unfold.

4 Dive Bomber 2

Bring sides and top down as you did for the Runway Skimmer.

5 Dive Bomber 2

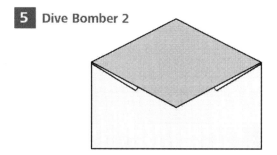

The completed basic fold.

6 Dive Bomber 2

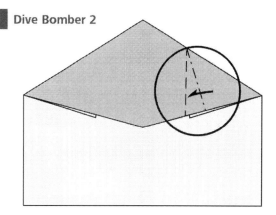

Larger view; fold the right-hand flap to the left, then open it out by tucking your finger in underneath, and then flatten the fold and repeat for the other side.

7 Dive Bomber 2

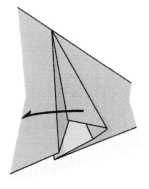

This image shows a cutaway where this fold from Step 6 is in progress. Flatten the fold; make sure it is symmetrical.

8 Dive Bomber 2

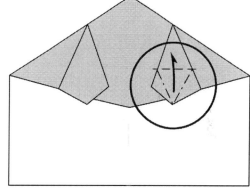

Now we will make the guns. Make creases on the right-hand flap, place finger underneath and lift the point upwards.

9 Dive Bomber 2

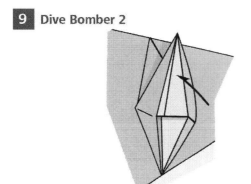

10 Dive Bomber 2

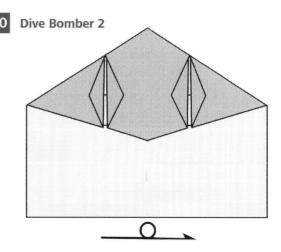

Step 9 cutaway shows the Step 8 instruction almost complete. Flatten sides to finish. Repeat for the other gun.

How it should look; now turn the model over.

11 Dive Bomber 2

12 Dive Bomber 2

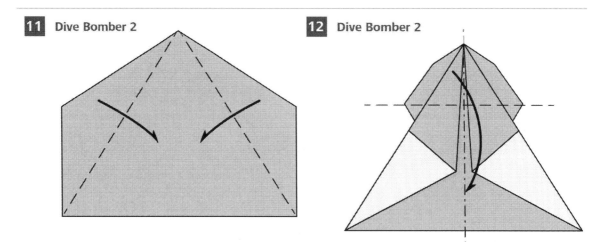

Fold the upper, outer flaps only.

Fold nose section back, making the guns swing forward from underneath, and then fold behind in half. Swing the model 90 degrees counter clockwise for Step 13.

13 Dive Bomber 2

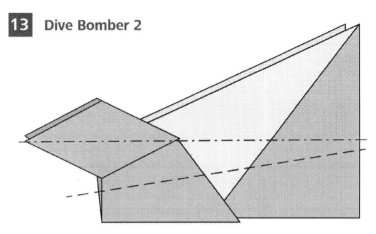

Make a horizontal mountain fold for the upper wing sections
and then fold down the entire wing/gun section.

14 Dive Bomber 2

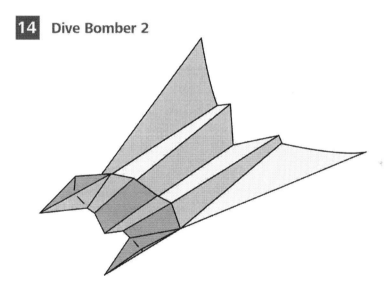

The completed Dive Bomber MKII is ready to fly. This craft can act as a regular glider, or diver, depending on lift (trailing edge tail section curled upwards) and method of throwing.

For dive bombing, drop at a steep angle, almost vertically. It will swoop down and gradually gain lift. Otherwise throw horizontally away from you for normal gliding. Invert fold a tail by half an inch if your craft dives too quickly (i.e. push inwards the bottom back corner at the end of the fuselage).

Difficult Folds in Detail

Here are more details on creating the base fold and guns for this model.

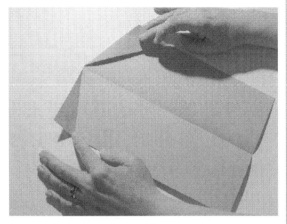

Having folded the corners in for a regular dart, and unfolded them. Fold the top corners in to meet the creases from the previous folds.

We will now make a similar base fold to that shown in the Super Looper instructions. Fold one diagonal; upper right folds across to align with left edge.

Unfold.

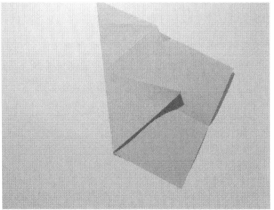

Repeat for the opposite diagonal and then unfold. Turn the paper over.

Make a horizontal fold that intersects the two diagonal folds you just made.

Unfold and turn the paper over.

Press your finger in the middle where the folds intersect. It will 'pop up'.

Bring the sides together.

Almost there; the sides should collapse inwards.

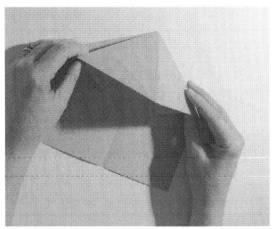

Flatten the top section.

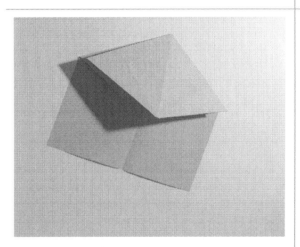

How it should look for the basic fold.

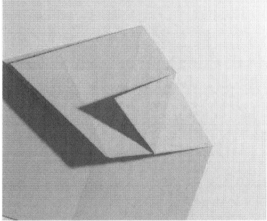

Now fold the upper-right flap to the left. Make the fold parallel to the center vertical crease; the leftmost edge will start where the underlying flap corners are.

Place your finger inside the flap and open it out.

Flatten the fold.

Crease-fold the lower corners of the new flap.

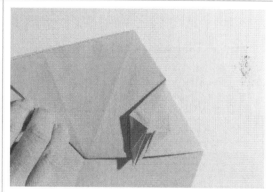

We will now unfold and lift the bottom edge of this flap upwards.

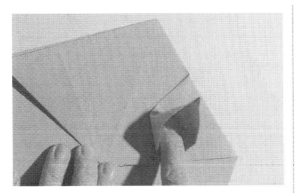

Almost there....

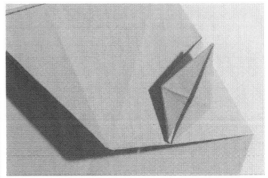

Flatten the sides and you have the gun section, and are ready to complete the glider at Step 10.

Some of the models in the rest of this book use similar base folds to that shown here.

Having mastered this craft, you should now have the skills necessary to make some of the more complicated models in the rest of this book.

Super Stunt Plane

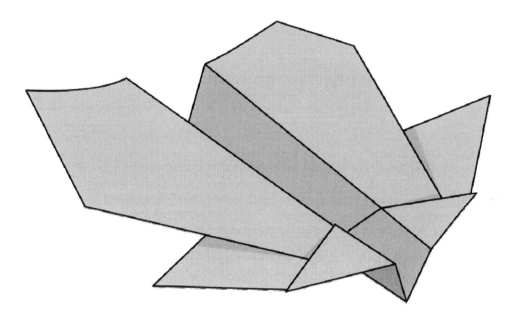

Similar to the Super Looper, this one does loops, circles and flies high to catch ocean breezes, or wherever there is wind. Great for throwing in open windy areas such as the park, beach, Congress etc.

This model uses the same basic fold as the Dive Bomber MKII. Have a look at the diagram in Step 1. Before you start at Step 1, make the usual center crease on a sheet of Letter or A4 paper, have corners folded in as for a regular dart, unfolded, and then folded to meet the new shallow-angle creases.

If you have trouble with the base fold, have a look at the Dive Bomber MKII *Difficult Folds in Detail* section for photos that provide more detail (shown on the preceding pages).

1 **Super Stunt Plane**

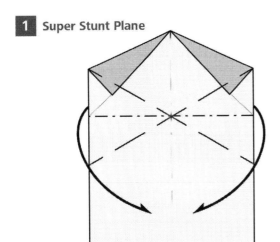

2 **Super Stunt Plane**

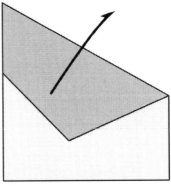

Make the intersecting diagonal creases (shown in the following three steps), plus a horizontal one that also intersects the diagonals. You will then be bringing the sides together to form a slightly different basic fold from the Super Looper.

First diagonal (crease fold/unfold).

3 **Super Stunt Plane**

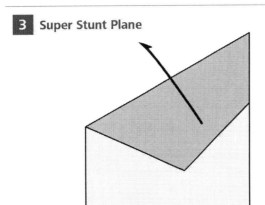

4 **Super Stunt Plane**

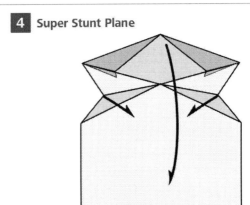

Second diagonal (crease fold/unfold).

Bring the sides together as you did for the Super Looper.

5 Super Stunt Plane

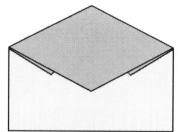

Unique basic fold complete! Phew!!

6 Super Stunt Plane

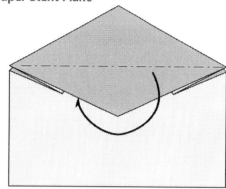

Larger view; tuck the lower point inside.

7 Super Stunt Plane

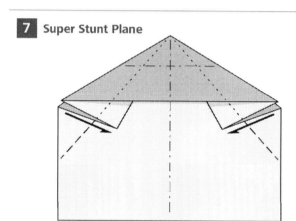

Fold lower sides in, under upper flaps. Next, fold the nose section behind, and then fold in half and swing the model about to face you for Step 8.

8 Super Stunt Plane

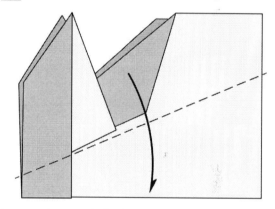

Larger view; tuck the lower point inside.

9 Super Stunt Plane

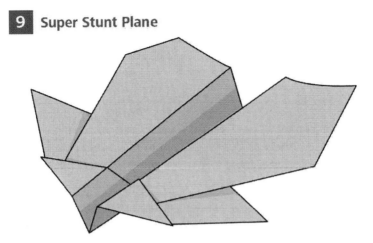

Completed craft; for loops throw vertically with force. For circles throw upwards and away from you with underside facing you. It should return for you to catch. To catch an upward draft, curl front wing-trails up slightly and then curl up the tail wing trailing edges.

Vertical Takeoff (Jump) Jet

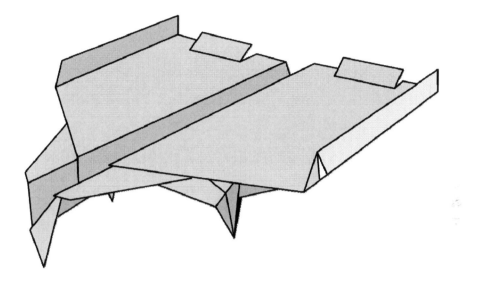

While it doesn't quite look the part of a sleeker jet, this craft does glide and land well on any surface as its steep angle of descent makes it suitable for landing on shorter runways...such as the aircraft carrier in the book, *Paper Boats!*.

1 **Vertical Takeoff Jet**

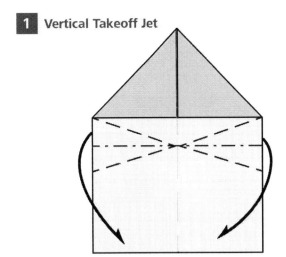

Begin with paper having the corners folded in as for a regular dart. Make creases and complete the basic fold (shown for your convenience in Steps 2 – 6). The horizontal dot-dash (mountain fold) line means you crease behind.

2 **Vertical Takeoff Jet**

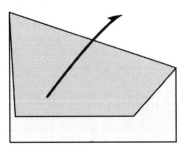

Shallow diagonal fold to left, then unfold.

3 **Vertical Takeoff Jet**

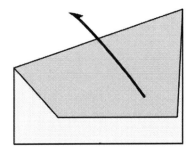

Shallow diagonal fold to right, then unfold.

4 **Vertical Takeoff Jet**

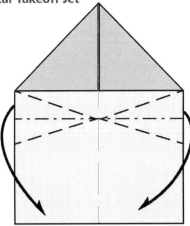

Don't forget the horizontal mountain fold. By pressing with your finger where the folds intersect, the paper should be able to come together easily.

5 Vertical Takeoff Jet

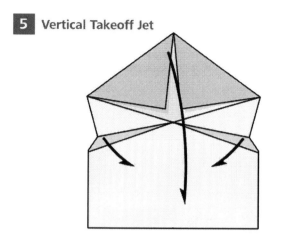

The sides and top should come together easily, just as you would have done for the Super Looper and other models in this book that use similar basic folds.

6 Vertical Takeoff Jet

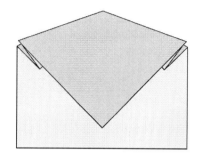

The finished basic fold.

7 Vertical Takeoff Jet

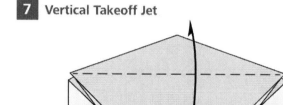

Larger view. Fold the flap upwards.

8 Vertical Takeoff Jet

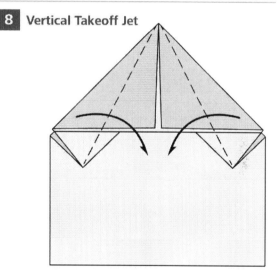

Fold the upper flaps inwards.

9 **Vertical Takeoff Jet**

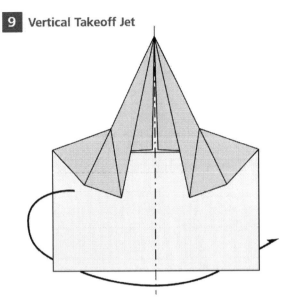

Fold in half behind.

10 **Vertical Takeoff Jet**

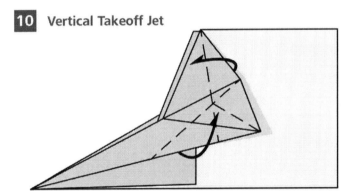

In Step 10, make the creases and pinch the edges together to form
a rabbit/bunny ear; this will become our undercarriage. The next steps
show cutaways of this undercarriage procedure in detail.

11 Vertical Takeoff Jet

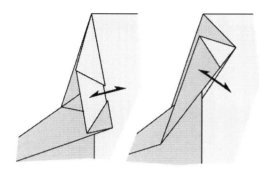

Make crease folds in both directions shown, and then pinch the sides together to make the rabbit ears.

12 Vertical Takeoff Jet

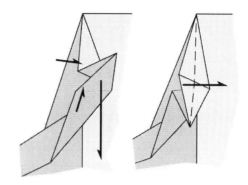

As you flatten the fold, swing the point down and in the right image, fold edge to the right to reinforce the undercarriage.

13 Vertical Takeoff Jet

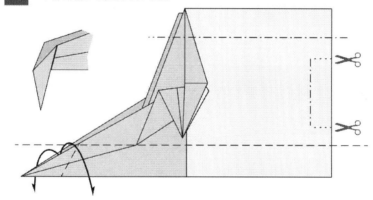

In Step 13, your undercarriage will look like this. All you need to do is fold the nose section: reverse-fold (inside out) to make a 90 degree-angle nose/forward landing leg, and then fold wings and fins, and finally cut trailing edge to add tail lift if necessary.

14 Vertical Takeoff Jet

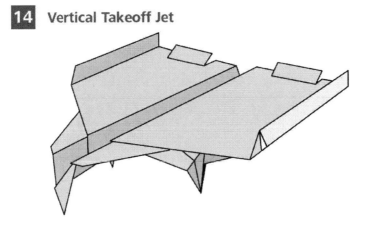

The completed jet; throw horizontally with a gentle motion. Do not launch at high angles as it will stall.

Modification Experiment

For a more sleeker-looking craft, it's possible to experiment with the wing span by doing a clever inverse fold at the top section of the undercarriage (before making the wing fins) and angling back the wing edge to the trailing wing tips. I'll leave the experimentation up to you, but here is my first attempt...

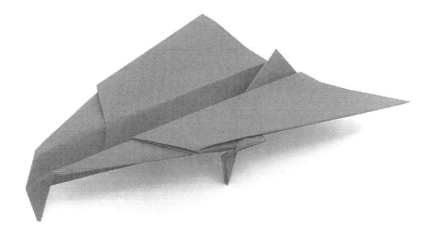

Modified jet

Concorde

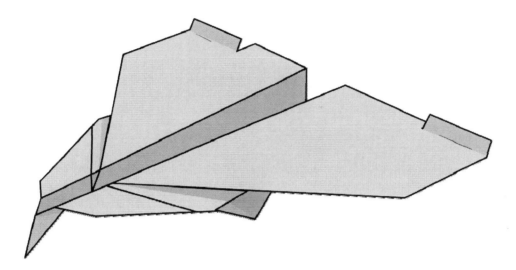

This graceful craft can once again streak across the skies and it's relatively easy to make, and noise-free! This model uses the same basic fold as the Vertical Takeoff Jet, the base fold steps repeated here for your convenience.

1 Concorde

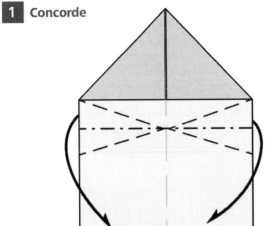

Begin with paper having the corners folded in as for a regular dart. Make creases and complete the basic fold (shown for your convenience in Steps 2 – 6). The horizontal dot-dash (mountain fold) line means you crease behind.

2 Concorde

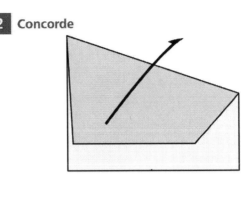

Shallow diagonal fold to left, then unfold.

3 Concorde

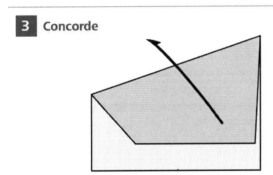

Shallow diagonal fold to right, then unfold.

4 Concorde

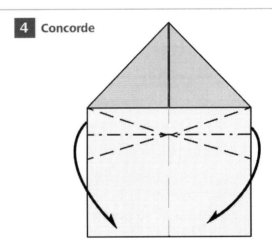

Don't forget the horizontal mountain fold. By pressing with your finger where the folds intersect, the paper should be able to come together easily.

5 Concorde

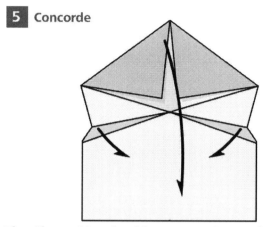

The sides and top should come together easily, just as you would have done for the Super Looper and other models in this book that use similar basic folds.

6 Concorde

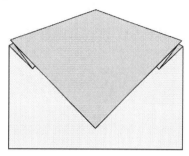

The finished basic fold. Note that depending on the paper used (A4 or Letter) the nose point may touch the bottom edge of the paper. A slightly different aspect ratio will not affect the performance of your craft.

7 Concorde

8 Concorde

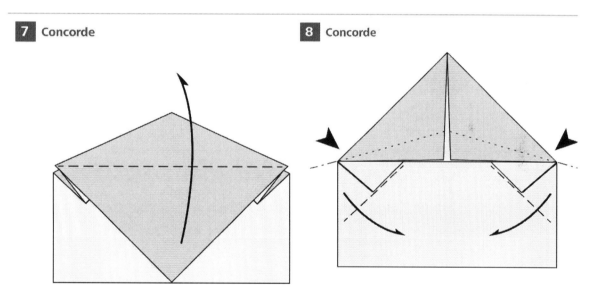

Larger view. Fold the flap upwards.

Push in the hidden edges (under top flap) by opening out the fold and bringing the sides inwards. Note the hidden edge; you are inverse-folding along this edge to complete the inverse fold. It is best to crease-well before starting this fold.

9 Concorde

10 Concorde

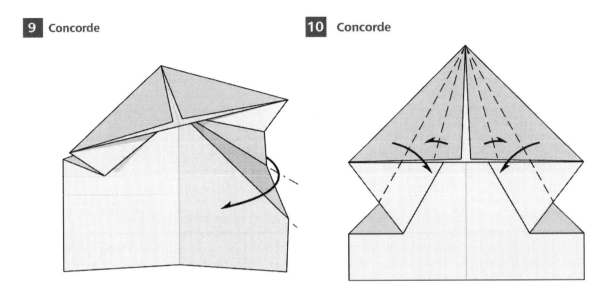

Almost there for the right-hand side... Repeat for the left side and flatten the fold.

Fold the outer sides inwards to make wing edges, and fold the inner flaps (closer to the center) underneath the outer folds.

11 Concorde

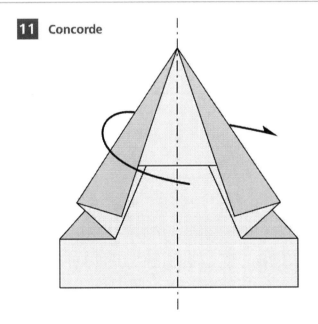

Fold in half behind.

12 Concorde

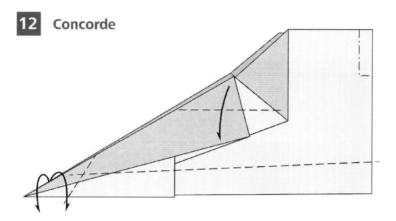

In Step 12, crease and open out the nose section, and reverse the fold (making it inside out). Next, fold down the simple landing gear, and wings, on both sides, and then add tail lift for stability by curling or snipping in by a quarter inch and folding rear lift.

13 Concorde

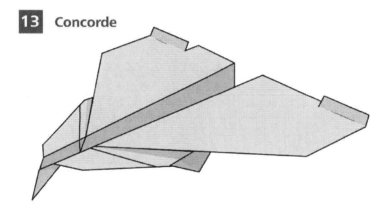

The completed Concorde; throw horizontally with a gentle forward motion. It will glide well before landing gracefully. There is a video of this on YouTube.

The following photo shows the Concorde parked in The Hangar (from *The Best Advanced Paper Aircraft Book 3*). There is also a modified version shown in that book.

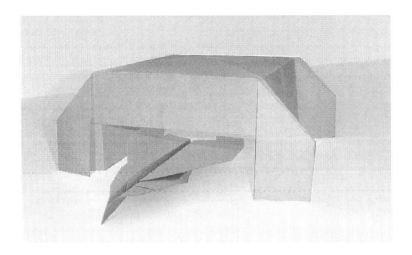

The Hangar

Winged Water Bomber

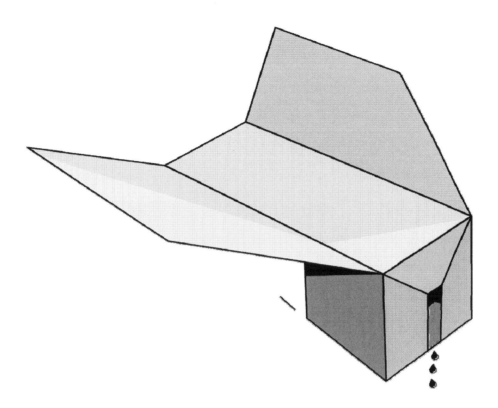

This flying projectile will certainly get the attention of your victims! While the flight path may be unusual, it sure delivers the message with a resounding 'splat'.

Before starting, you will need to change the aspect ratio by snipping off an inch or more from the long side of a piece of A4 or Letter paper. This will ensure you have enough paper for the wing section. Experiment with different lengths of paper to see how your bomber flies.

1 Water Bomber

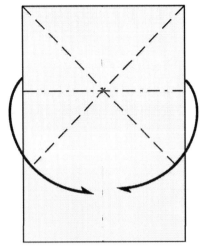

Funnily enough, we will make a basic fold, otherwise known as a 'water bomb base'. Make the two diagonal creases and a horizontal one behind that intersects the diagonals.

2 Water Bomber

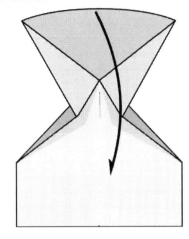

Bring the sides together. You should be used to this by now :)

3 Water Bomber

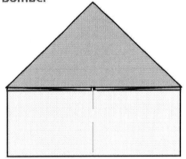

The completed basic fold.

4 Water Bomber

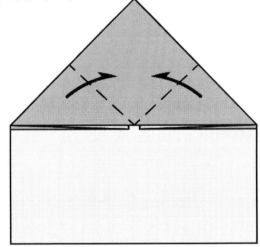

Larger view; fold the upper corners in to the center.

5 Water Bomber

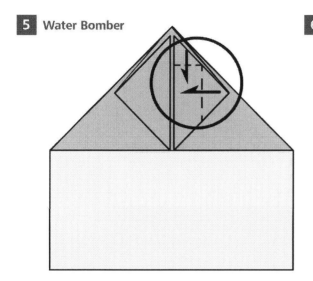

Lock the folds together, as for a regular water bomb.

6 Water Bomber

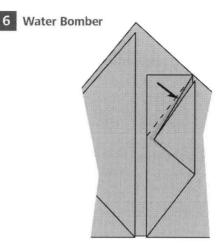

Steps 6 and 7 show the procedure for Step 5; tuck the corner into the triangle.

7 Water Bomber

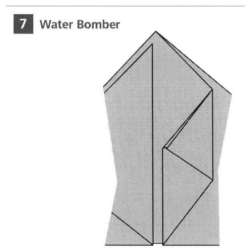

Repeat for the left side.

8 Water Bomber

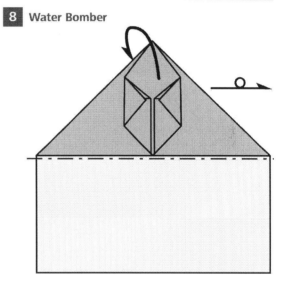

Fold the entire upper section behind, and then turn the model over.

9 Water Bomber

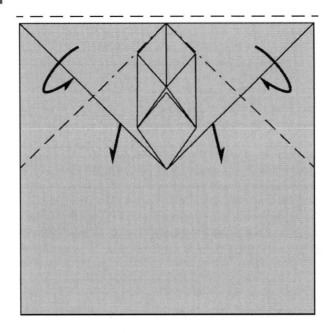

Crease-well both diagonals and invert the corners from the leading (top) edge.

10 Water Bomber

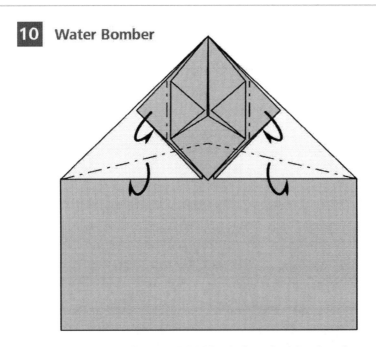

Tuck behind/inside the lower flaps and fold back the other two bomb sections to lock it.

11 Water Bomber

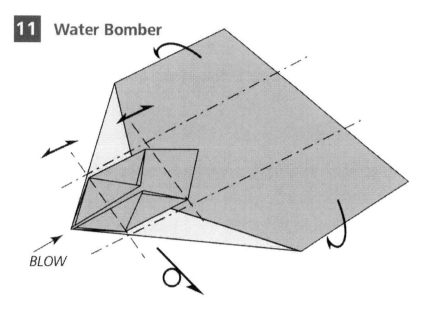

BLOW

In Step 11, crease-well the bomb section sides, as you would for a regular water bomb, and then fold back the wings parallel to the bomb section.

While carefully holding the wings at the leading edges, blow into the water bomb section.

12 Water Bomber

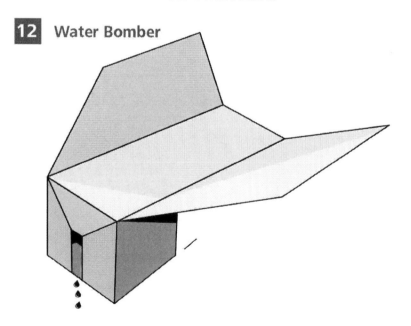

Your completed Water Bomber.

Fill with water (of course) or perhaps shaving cream, wet clay, or that sticky raspberry cordial, or JELL-O, or anything you like for maximum effect.

Throw hard in the direction of your victim. Make sure you capture this scene on your cell phone!

Experiment with the wings for directional tests. In an unfilled state, this craft acts as a cute spinner when thrown from a tall building, but somehow filling it with water has more satisfaction.

Long Nose Glider

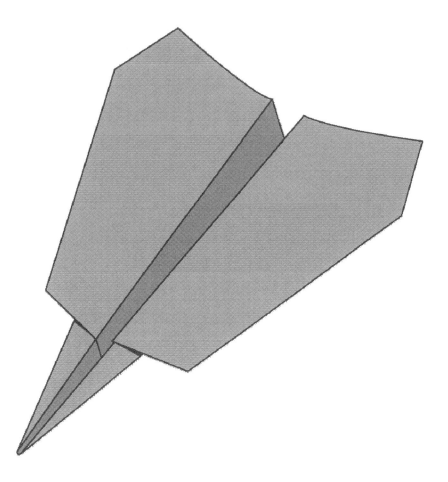

One of these became lodged above the boardroom door at our office. As the manager opened the door the plane flew inside and landed on the boardroom desk. Those inside were not amused. The rest of us were laughing over this event for a week!

1 Long Nose Glider

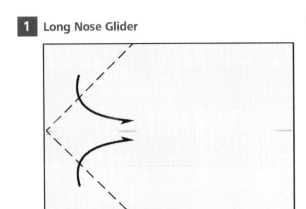

Fold the corners in.

2 Long Nose Glider

Fold again as for a regular dart.

3 Long Nose Glider

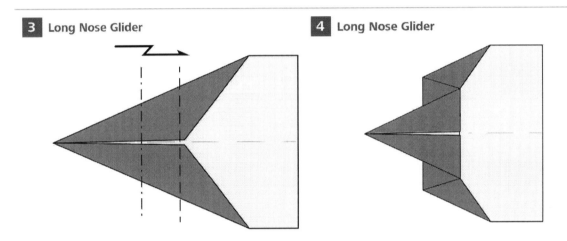

Make a 'stair-step' fold; valley fold the dashed line to the right and then fold back along the dot-dash line to the left.

4 Long Nose Glider

Your fold will look like this.

5 Long Nose Glider

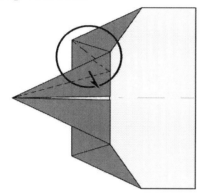

In the circled area we will narrow the nose. First make creases; your fingers will go under the upper left flap as you fold down the angled edge.

6 Long Nose Glider

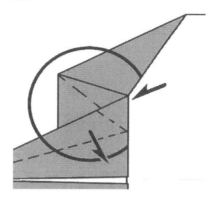

In Step 6, we're taking a closer look; as you fold down the angled edge of the nose, also fold down the upper corner edges where the right-hand arrow is pointing. It will flatten out as the lower edge is folded down.

7 Long Nose Glider

Step 7 shows a close-up cutaway of the fold almost completed. Flatten and then repeat for the opposite side.

8 Long Nose Glider

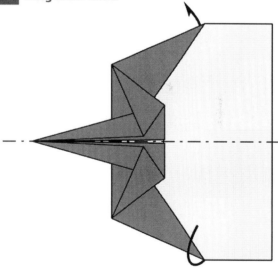

Fold behind in half.

9 Long Nose Glider

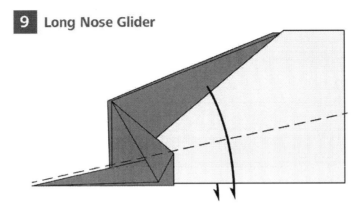

Fold the wings down along the angled nose axis, and you are then ready to fly.

10 Long Nose Glider

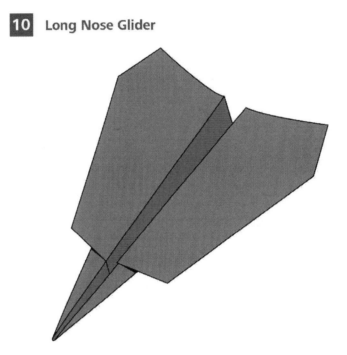

Throw high with moderate force. Note that if you throw it too hard it may stall. If you are throwing this model around people, fold back the nose point about half an inch to avoid sticking in an ear or eye, someone's hair, or up a nostril, or in a key lime pie.

This craft is great for throwing in large areas such as assembly halls, malls, court rooms, Congress, or the most elegant of upper-class restaurants – especially when there are key lime pies being served.

Winged Wonder

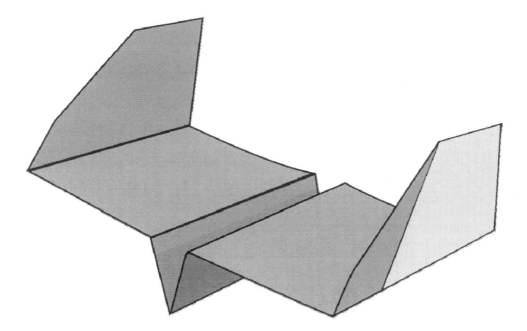

This is a simple yet agile craft that flies well in high wind conditions, staying aloft for longer, given the larger surface area wings, with no forward section to weigh it down.

1 Winged Wonder

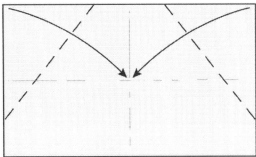

First start with paper that is folded in half both ways to get a center crease point, and then fold the top corners in to meet the center where the creases intersect.

2 Winged Wonder

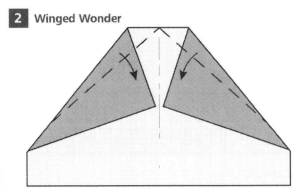

Fold the leading edges in.

3 Winged Wonder

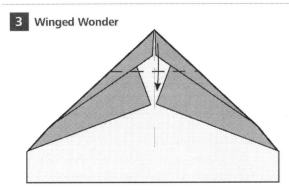

Fold the nose point down to where the two lower points meet.

4 Winged Wonder

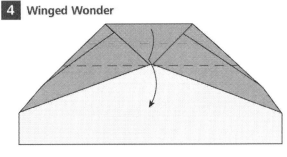

First fold the leading edge to where the points meet, and then fold again.

5 Winged Wonder

6 Winged Wonder

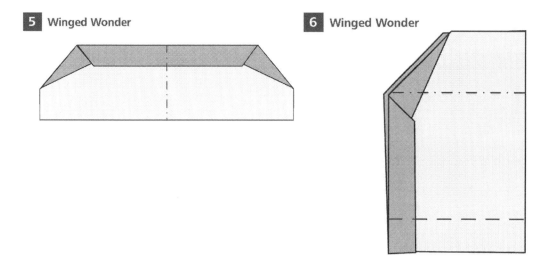

Fold in half behind.

Fold down the wings and fold back the large side fins.

7 Winged Wonder

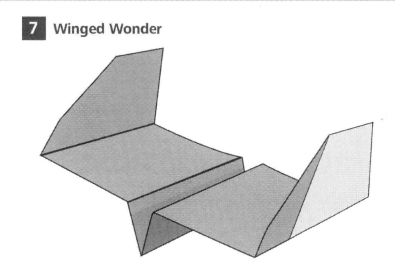

The wing is complete. Note that your model may look different if you are using paper having a different aspect ratio.

Gently throw forward in a slight downward angle for a smooth glide. Add some tail lift and throw high for loops or away from you with underside facing you for circles.

Try using A3 with two inches removed from the long edge to change the aspect ratio. Different proportions result in different side-fin dimensions as seen in the diagram, and therefore different performance characteristics.

If you throw harder and upwards and your craft suddenly dives, add more tail lift and/or fold leading edge of each side fin towards the center fuselage by a 10 degree angle. Glue the fuselage together if you want to throw with more thrust. Also, experiment by swinging the sides downwards and see what happens. :)

Deluxe Glider

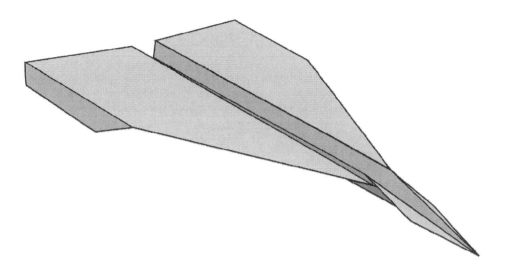

This is a smooth, elegant glider that loves breezes, upward drafts and long, leisurely flights. I threw one at a friend's beachside residence at Seacliff, California and the wind caught it and carried it away till it could no longer be seen, even with binoculars!

1 Deluxe Glider

2 Deluxe Glider

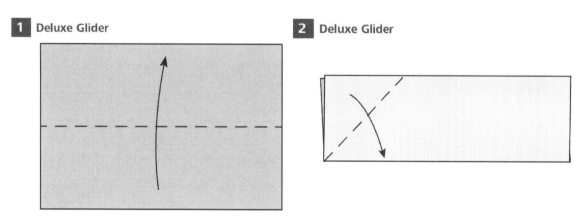

Fold lengthwise in half.

Fold corner down on both sides.

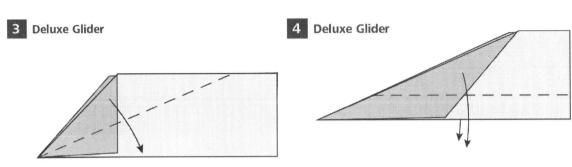

3 Deluxe Glider

Fold corner again on both sides.

4 Deluxe Glider

Fold down wings parallel to fuselage.

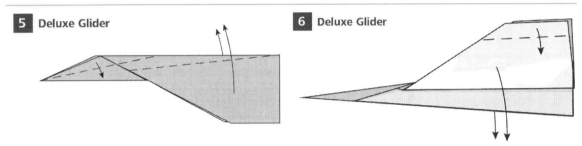

5 Deluxe Glider

Fold nose section and then fold up the wings at angle indicated.

6 Deluxe Glider

Fold the side fins down and then swing down the wings at 90 degrees to the fuselage.

7 Deluxe Glider

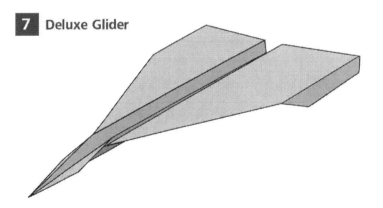

The glider is complete. Curl up the trailing edge for lift if necessary. Throw with moderate force with a horizontal or slight upward motion.

This craft is great for throwing off a cliff or tall building (like we do at Century City at the Plaza); it will fly some distance. If you are throwing near people, fold the nose back by a half inch or so; it should not affect performance.

Super Stunt Circler

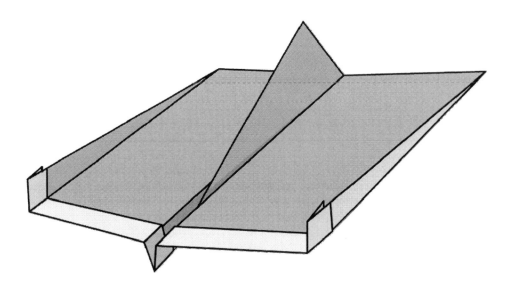

This craft can do some amazing stunts. No matter how I threw it, it kept returning. At a paper plane demonstration in our local mall, I threw this craft twenty times and each time it returned to my hands unfailingly.

Sometimes on windy days it would catch a breeze and be carried away so it's a good glider too.

It is recommended to modify an A4/Letter or A3 sheet of paper to get a longer aspect ratio, proportionally to the shape shown in Step 1 (cut an inch off the long side). But you don't have to; try Letter first, and then try snipping an inch off the long side to experiment in circling behavior if the craft comes up short.

1 Super Stunt Circler

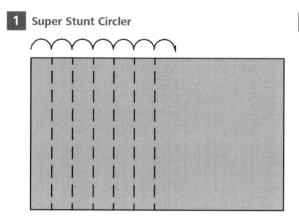

Fold the left edge over by six times, approximately resulting in a leading folded edge mid-way. The fold can be around ¾ inch each time. This gives you a solid leading edge wing.

2 Super Stunt Circler

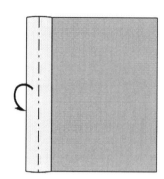

Fold the leading edge in half behind.

3 Super Stunt Circler

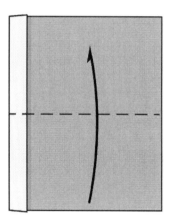

Fold in half and crease well for the leading edge; make sure there are no kinks/deformities.

4 Super Stunt Circler

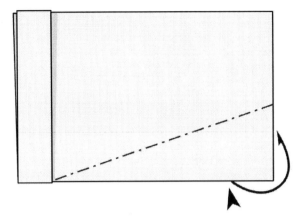

Crease well and push in (invert fold) to make the tail.

5 Super Stunt Circler

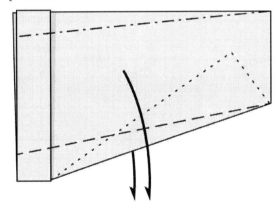

Fold down wings and fins parallel to each other. Crease well the leading edge that becomes the fuselage that you will hold to throw the model.

6 Super Stunt Circler

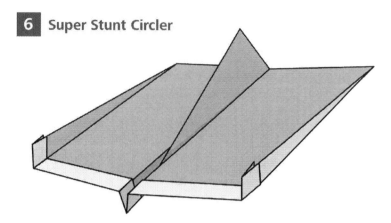

Completed aircraft; for a loop throw hard and high at 45 degrees; it will reach its peak and arc back to you. For circles, have the underside facing you and throw with moderate force away from you at a 20–30 degree arc.

Try curling up the tail section, one at a time, if the craft proves recalcitrant, or experiment with different lengths of paper. Once you master this craft, it will faithfully return to your hand, time and time again.

Recon Glider Surveyor

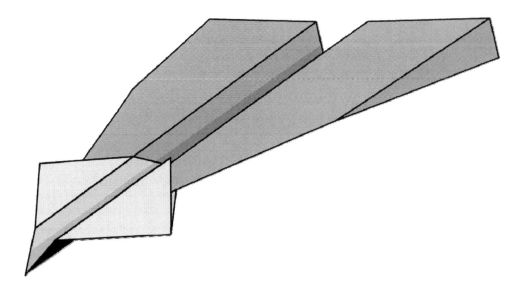

My kids love making this glider as it's so versatile for any situation; be it at college, lecture halls, the cinema, the football/baseball stadium, Congress or wherever there is fun to be had (only kidding!).

This model has a heavier forward section than a regular dart, so you can throw it hard. Its wingspan will allow it to glide a good distance.

1 Recon Glider Surveyor

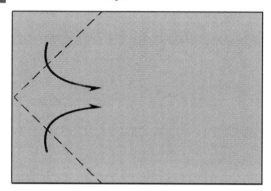

Fold corners.

2 Recon Glider Surveyor

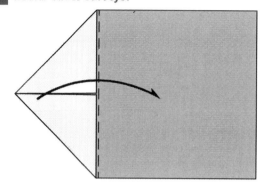

Fold point to the right.

3 Recon Glider Surveyor

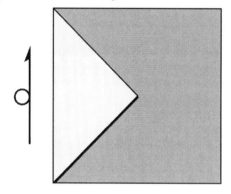

Turn the paper over.

4 Recon Glider Surveyor

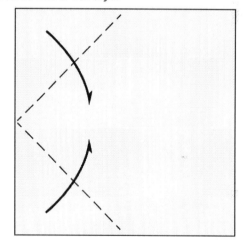

Larger view; crease well when you fold the corners in again.

5 Recon Glider Surveyor

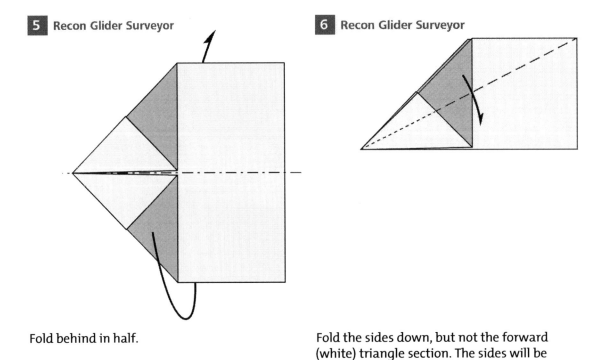

Fold behind in half.

6 Recon Glider Surveyor

Fold the sides down, but not the forward (white) triangle section. The sides will be folded down and inside the triangle section. Step 7 shows this near-complete.

7 Recon Glider Surveyor

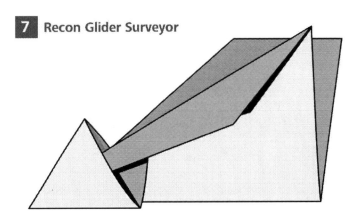

Flatten the fold and repeat for the opposite side.

8 Recon Glider Surveyor

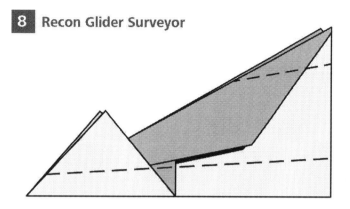

Fold wings and fins down at a shallow angle.

9 Recon Glider Surveyor

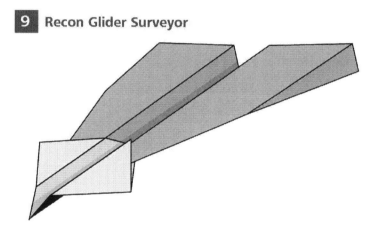

Throw steadily with moderate force. The solid nose section can take considerable thrust, so experiment by throwing harder for longer flight times. Also try small triangle side fins on the forward section (see photo). This craft flies level for a fair distance but may curve to the right or left, so adjust tail lift slightly on either trailing edge to correct.

Recon Stunt Tactical

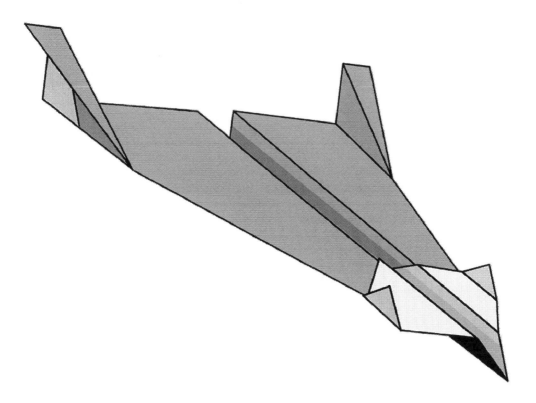

This is a variation on the Surveyor model with tail lift for huge looping behavior. We have actually attached this version on top of the surveyor model and watched them separate mid-flight. Perfect for high-altitude A-12 OXCART-style surveillance, and a separation module when you need it (to escape those Russian missiles!).

Begin with Step 8 of the Recon Glider Surveyor and note the different folds for the wings, as follows...

8 Recon Stunt Tactical

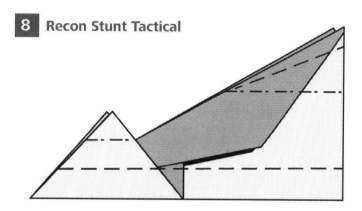

Fold wings down parallel to the bottom of the fuselage. Fold up the small triangle fins in the forward section. Fold tail fins up instead of down, and fold back the leading edge of each tail fin. This will allow extra lift for looping.

9 Recon Stunt Tactical

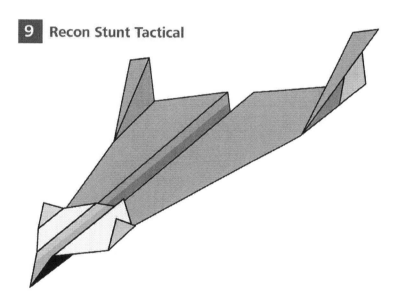

Aim high and throw hard for some huge loops. Outdoors is best for this, or a big mall. Add tail lift if necessary. Experiment with the different folds for the side fins (up/down) and see what happens.

Lightwing

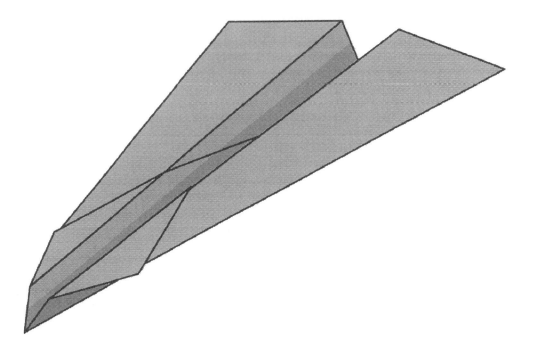

This craft features a wing of one paper thickness where most of the folds have gone into the forward section and fuselage, meaning you can fling this craft hard and it will glide for a considerable distance.

1 Lightwing

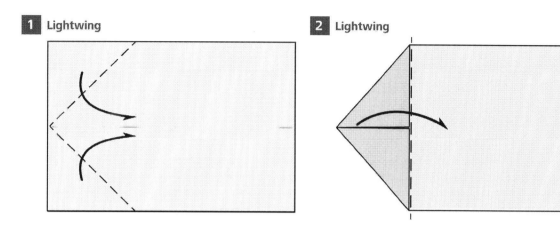

2 Lightwing

Fold corners as for a regular dart.

Fold point to the right.

3 Lightwing

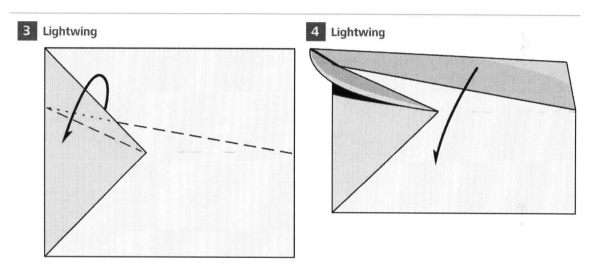

4 Lightwing

Larger view; tuck your finger under the upper left flap and fold this angled edge to meet the center crease, while folding the rest of the side down. Do no overstep the center crease.

Almost there....

5 Lightwing

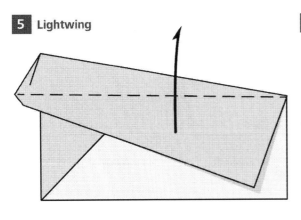

Fold back this flap along the center crease and repeat Steps 3, 4, and 5 for the opposite side.

6 Lightwing

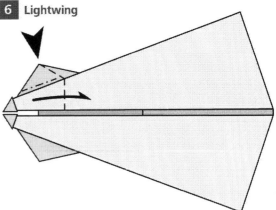

Crease this section well and push the point down while swinging it to the right.

7 Lightwing

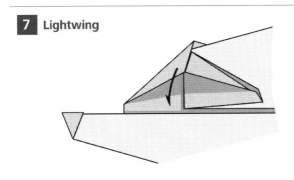

Step 7 shows cutaway of the step 6 task almost complete; flatten the fold and repeat for the opposite side.

8 Lightwing

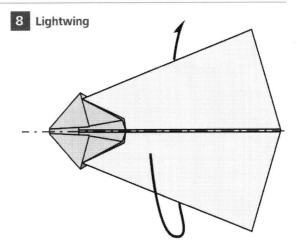

Fold in half behind.

9 Lightwing

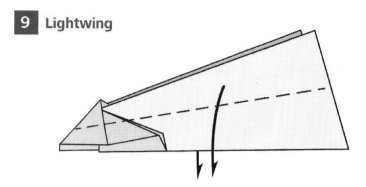

When you fold down the wings, crease the forward section well.

10 Lightwing

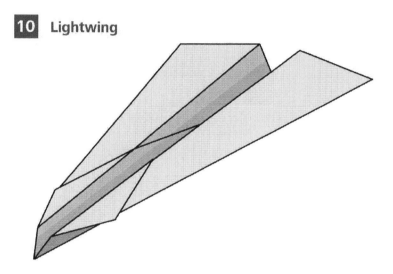

You are ready to fly! Throw hard and at 20–30 degrees. Add slight tail lift by curling up the trailing edge. You can throw this model very fast and meet less air resistance than a regular dart. This craft is great for speed competitions.

Sleek Glider

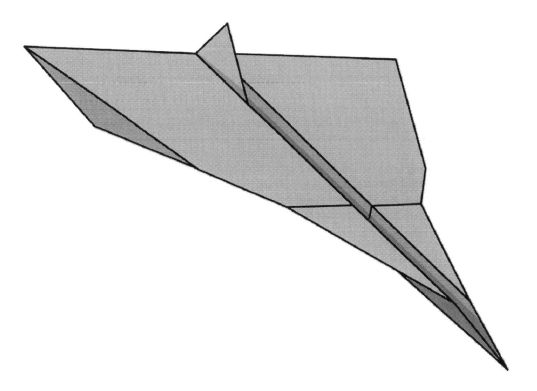

This is a nice-looking glider that is not too difficult to fold and glides smooth and level.

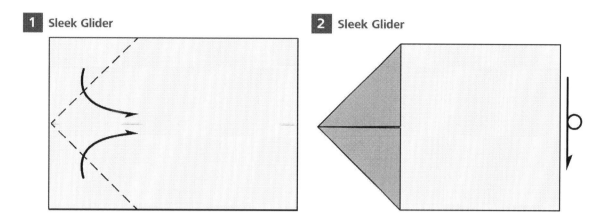

1 Sleek Glider

2 Sleek Glider

Make sure paper has a horizontal center crease. Fold the corners in.

Turn the paper over.

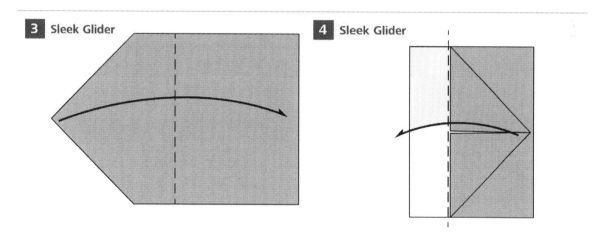

3 Sleek Glider

4 Sleek Glider

Fold left point across to right-hand edge.

Fold the right-hand upper section to the left.

5 Sleek Glider

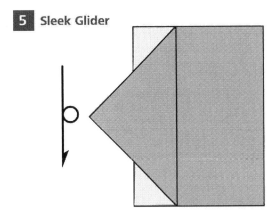

Turn the paper over.

6 Sleek Glider

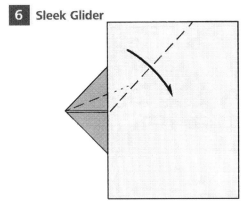

Fold the upper vertical edge to meet the center horizontal crease, while folding the nose section back at the same time. The larger view in Step 7 shows this action almost complete.

7 Sleek Glider

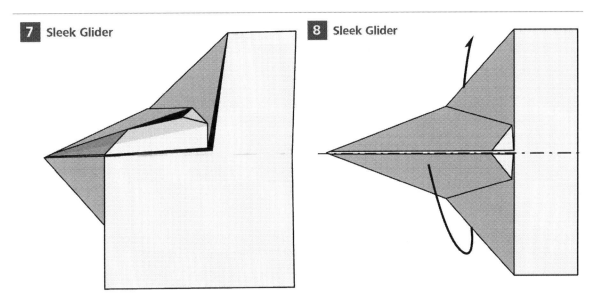

Almost there... Repeat Step 6 and 7 for the opposite side.

8 Sleek Glider

Fold behind in half.

9 Sleek Glider

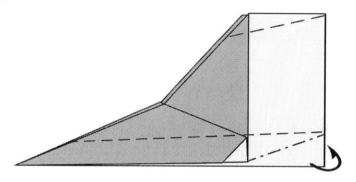

Crease the bottom right corner tail section and then push inwards (invert fold).
Next, fold down wings and fins.

10 Sleek Glider

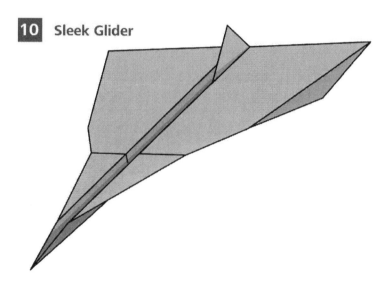

Throw gently away from you for easy gliding. Throw harder for some interesting aerial manoeuvres. You may wish to seal any loose folds under the fuselage or even seal the fuselage itself with tape so you can throw your craft faster. If the nose gets crumpled on impact, fold back by half an inch; it should not affect performance.

Ace Flier

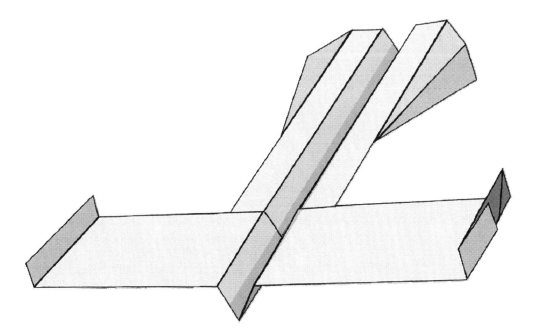

Found the sketches for this craft while rummaging through my old notes, a craft I designed when 14. I knew there had to be a way to make something like this with one piece of paper instead of two like my school pals were doing.

This craft can do giant loops, circles, or just plain hitch a ride on a gust of wind. Nice. It has a strong fuselage, meaning you can throw it hard and high against any wind with less chance of the fold coming apart. This craft has appeared in many apps — and finally in my own *Paper Aircraft Advanced* app :)

If you use paper colored one side, white the other, having the color underside means you can spot the airplane better from underneath, and follow it as it flies away – because this model often gets carried away on a breeze.

It's recommended to trim an inch off the long edge of a sheet of Letter paper, though this may not be necessary, depending on how well you folded it. Experiment :)

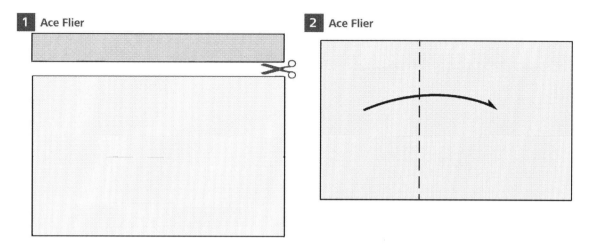

1 Ace Flier

Trim an inch off the long side.

2 Ace Flier

Fold in relative position shown (do not fold in half).

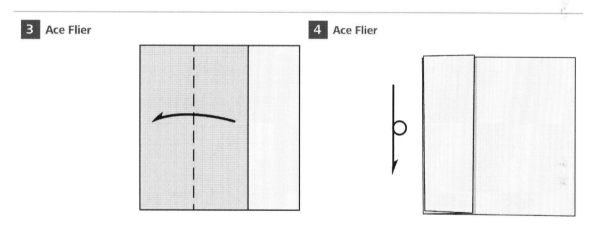

3 Ace Flier

Fold upper flap to left.

4 Ace Flier

Turn the paper over.

5 Ace Flier

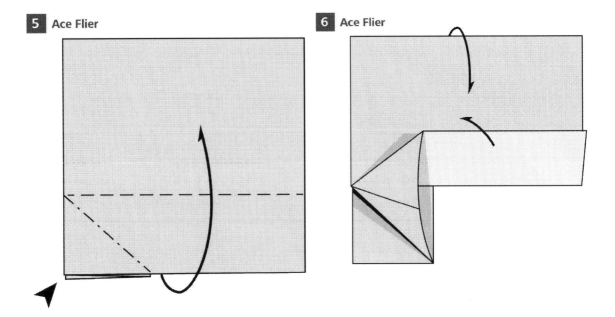

With the model turned over, carefully lift the bottom edge, with the bottom left corner opening out. Best to crease fold the bottom left corner first.

6 Ace Flier

Almost complete; repeat for the other side.

7 Ace Flier

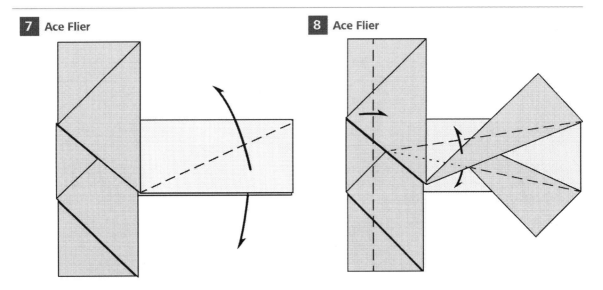

Fold tail wings out.

8 Ace Flier

Fold fuselage sections for reinforcement, and then fold leading edge wing.

9 **Ace Flier**

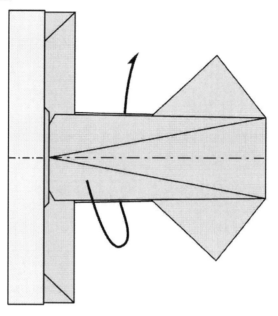

Fold in half behind.

10 **Ace Flier**

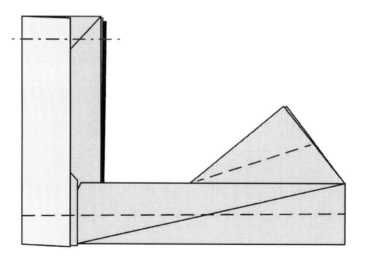

Fold down wings parallel to fuselage, fold angled tail wings
and fold back the side fins.

11 **Ace Flier**

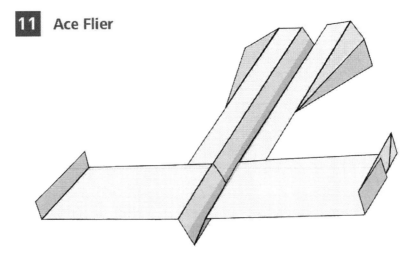

Complete. Phew!

Throw upwards at a 45 degree angle for loops. Throw with force as the wing strength will be able to handle more thrust. Try different throwing angles for loops, circles and catching upper winds.

Loop Glider

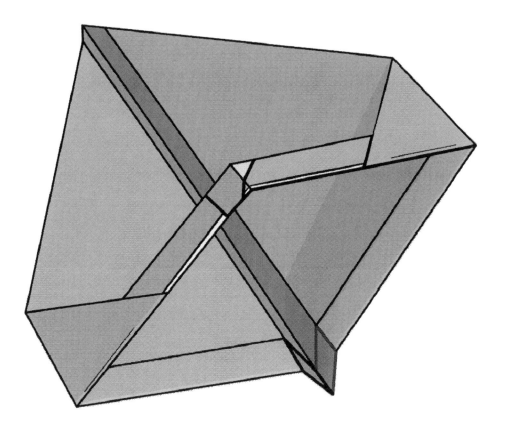

An interesting craft that confuses some people: when they pick it up they tend to throw it end first. Don't let the directional inference fool you, this craft is thrown by the heavy looped forward section, and it glides amazingly well.

1 Loop Glider

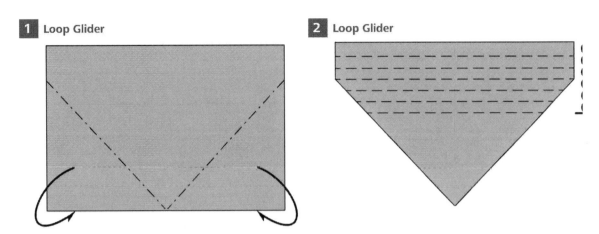

Fold up the bottom corners behind (or you could fold the other way around and then turn over).

2 Loop Glider

Fold down the leading edge six times at just under half inch intervals approximately.

3 Loop Glider

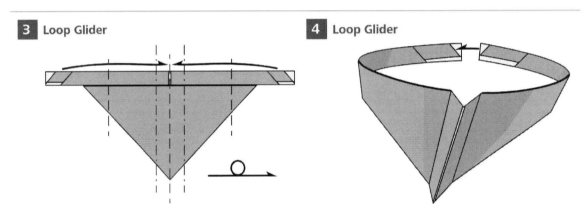

Fold the fuselage section and angled wing sections. We will lock the forward section folds together...

4 Loop Glider

Slot one end into the other. Step 5 shows a close-up cutaway.

5 Loop Glider

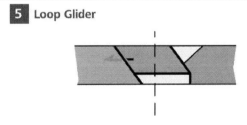

Complete the locking action by sliding one end into the other by an inch or so. Once you have connected the ends together, fold in the middle.

6 **Loop Glider**

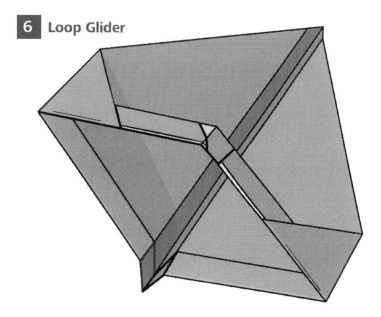

Finished glider; to throw, hold the fuselage with finger and thumb and throw high with moderate force. It may first fly erratically; therefore adjust the looped section and experiment.

Try inverting the tail section to make a small fin for better stability. Lengthen or shorten the leading wing edge slot to determine the best performance. This model flies well using larger paper too.

Flight Theory and Throwing Skills

To understand how your plane flies, how it behaves mid-flight, is important in making your plane fly well. There are four basic forces that control the flight of a paper plane (or any plane for that matter). They are:

- Lift
- Drag
- Gravity
- Thrust

Lift is a force that pushes an aircraft upwards against the force of gravity.

Drag is the normal force of air opposing an aircraft's forward motion.

Gravity is a natural force that pulls an aircraft to the ground. Thrust is a force that propels an aircraft through the air, opposing drag.

Thrust is created by an aircraft's propeller, jet engine or in the case of paper aircraft by the motion of your arm throwing the plane.

In the following diagram, the kinetic energy of thrust from your arm propels the glider in the direction (the intended glide path) you wish it to go. It is met by air resistance (drag) and the glider's lift is temporarily greater than gravity, making it climb.

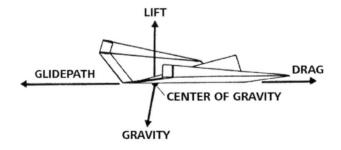

As the kinetic energy expires, gravity overcomes lift and your plane glides to the ground.

Lift is created by a drop in air pressure above an airplane's wing. A typical airplane's wing has a curved upper surface so that while air moves over the wing its speed increases and causes a pressure drop. The air under the wing remains at normal pressure so the wing can only go upwards.

Stability and Control

The stability and control of an airplane is another factor well worth noting. An airplane has three basic movements:

- Yaw
- Roll
- Pitch

A plane makes each movement on an imaginary axis that passes through the center of gravity. In the following figure, we see that Yaw is a plane's movement on its vertical axis as its nose turns left or right.

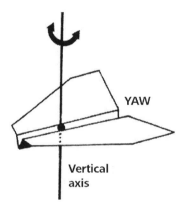

In the following figure, we see that Roll is a plane's movement on its longitudinal axis as one wing tip drops lower than the other.

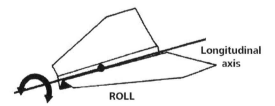

In the following figure we see that Pitch is a plane's movement on its lateral axis as the nose moves upwards or downwards.

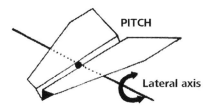

Aircraft Modifications

A paper airplane can have many controls to allow it to perform the way you want it to. Some of these are tail fins, wing fins, rudders, elevators (rear wing lift), ailerons (forward wing lift) and a whole lot more, including rear inverted tail loops for lift and stability (at the expense of more drag).

Such devices are used by many aircraft designers and I encourage you to experiment with your craft to see how it behaves after a modification.

Throwing That Perfect Plane

Every paper plane has its own 'personality'. This is largely due to the increasing variable state from the initial condition of the launch moment. A craft can be recalcitrant despite being made beautifully, it can be graceful but fall short, it can have that peculiar kink in its glide path that sends it barrelling, or teetering to one side (remember 'symmetry'), or it can be just plain stubborn, stalling at the slightest breeze. These of course are all the laws of nature than can occur after our initial condition, and despite our best efforts, chaos always remains due to variables we cannot control; a slightly stronger breeze than before, enough to topple our craft, or a bug crossing our flight path.

However, we reduce the potential size of our initial condition by using the correct paper, with practice in folding, and throwing. In other words, the greatly varying results of a series of the same kind of throw can be reduced by symmetry and good throwing practice. Assessing potential variances on trajectories prior to launching can also help, and this is discussed later in this guide.

Putting our light touch on physics aside, feeling confident and knowing how to throw your craft is the essential complementary skill to building it in the first place.

Let's get down to details...

Darts

Darts made from good stiff paper can travel a long distance. Typically hold the dart in the center of the fuselage, aim 20 degrees above the horizon and thrust your arm in a forward motion, releasing the craft as you feel the lift under the wings.

As you launch, while holding the dart, keep wings level until the point of release. Throwing with some force and keeping your arm straight after release is recommended, mainly because if you move your arm away too quickly, you could create a downward shear, or accidentally touch the empennage, the trailing edge of your craft, causing it to change trajectory or even stall.

Treat your launch as if you were taking a golf swing. Don't let anything distract you and keep your target in mind when in the motion of throwing your craft.

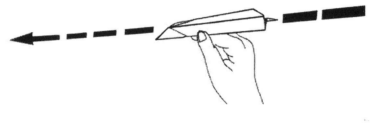

Gliders

With wing spans that are typically wider than darts, gliders call for a more gentle forward motion from your arm. Aim 10 to 20 degrees above horizon. For craft with shorter fuselage, aim at the lower angle of 10 degrees, otherwise unwanted lift will raise short-winged craft too early and stall them.

You do not need to consider the throwing angle so much if you are throwing a glider from a cliff top or tall building, in which case a more gentle motion with your arm is called for. Such heights are often accompanied by upward wind pressure. You want your craft to be carried on this wind, rather than be swept back towards you. Throw gently, as if you are throwing the craft onto another carrier, the air, which will take your glider further.

Stunt Gliders

These 'multi-talented gliders' can glide, loop and do circles. Few gliders have all these traits but some do (a few in the *Paper Aircraft Advanced* app and *Best Advanced Paper Aircraft* series work well).

For loops, hold the middle of fuselage (slightly more toward front wing for front-winged craft) and thrust high into the air between 40 and 50 degrees. Add a quarter inch curl to the wingtips if necessary to ensure a complete loop. Sometimes wind direction can alter a plane's looping performance. Experiment by throwing directly against the wind as well as side-on, and you will soon become accustomed to the glide behavior.

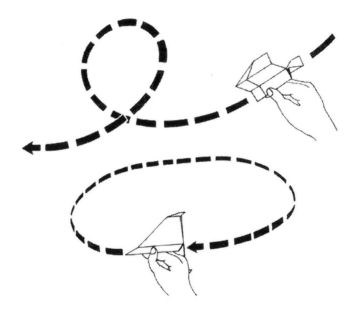

Circles are trickier; hold the center of the fuselage with the underside of the craft facing you, parallel to your body. Tilt the craft up 15–25 degrees and throw at a 15–25 degree arc away from you. The craft should swing back so that you can catch it. Right-handers, throw left. Left-handers, throw right. A slight curl on one trailing wing tip or the other will improve the turning circle.

Odd-shaped Gliders

The best advice for unusually-shaped gliders is to experiment. Various-shaped gliders in my other books are purpose-designed to cheat windy days, to spin, 'heliglide', and so on.

If you're designing an unusual craft, experiment by just dropping the craft from a height and see how it behaves in the air without thrust. If it starts to spiral, correct the wings; or enhance that aspect and make it a true spinner!

Paper Aircraft Advanced App Now Available

  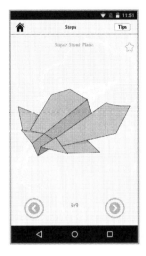 

Download this free app to your iPhone or Android device. Features ten free models for go-anywhere folding, and has no annoying ads. Up to 34 models are available.

iPhone:

Android:

iPhone:

https://itunes.apple.com/WebObjects/MZStore.woa/wa/viewSoftware?id=986661964&mt=8

Android:

https://play.google.com/store/apps/details?id=com.paperaircraft

Printed in Great Britain
by Amazon